Technische Universität Dresden

Dissertation

Vertical Gallium Nitride Power Devices: Fabrication and Characterisation

M.Sc.

Rico Hentschel

der Fakultät Elektrotechnik und Informationstechnik der Technischen Universität Dresden

zur Erlangung des akademischen Grades

Doktor der Ingenieurwissenschaften

(Dr.-Ing.)

Dresden 2020

ISBN: 978-3-7526-4176-9

Herstellung und Verlag: BoD- Books on Demand, Norderstedt

Abstract

Efficient power conversion is essential to face the continuously increasing energy consumption of our society. GaN based vertical power field effect transistors (FETs) provide excellent performance figures for power-conversion switches, due to their capability of handling high voltages and current densities with very low area consumption. This work focuses on a vertical trench gate metal oxide semiconductor field effect transistor (MOSFET) with conceptional advantages in a device fabrication preceded GaN epitaxy and enhancement mode characteristics. The functional layer stack comprises from the bottom an n^+/n^--drift/p-body /n^+-source GaN layer sequence. The layers up to the p-GaN were grown by metal organic vapour phase epitaxy (MOVPE), whereas – special to this thesis – the topmost n^+ layer was grown by molecular beam epitaxy (MBE) after a hydrogen out-diffusion anneal. This last growth step in a hydrogen-free environment prevents re-passivation of Mg-doping agents in the p-GaN by hydrogen.

Fabrication specific challenges of the concept are related to the complex integration, formation of ohmic contacts to the functional layers, the specific implementation and processing scheme of the gate trench module and the lateral edge termination of the MOSFET. Firstly, cost efficient pseudo-vertical MOSFETs were utilized. These devices are very suitable test vehicles since they provide a fast feedback loop between growth development and device processing. In a second step, bulk GaN substrate based true-vertical devices with thicker drift layer are fabricated. Cooperative test structures allow the micro-structural and electrical characterisation of the different device building blocks and thereby provide essential information and a better understanding of the functionality of the entire MOSFET. Important test structures incorporate e.g. the vertical $TiN/Al_2O_3/GaN$ gate stack or the pn^- body diode.

As material quality has huge impact on the final device performance, the quality of growth and overgrowth of the active layers is studied by mapping the surface morphology with atomic force microscopy (AFM). Endpoints of threading dislocations on the surface are decorated by defects, which enable an estimation of the dislocation density. It is found that the type of decoration as pits or hillocks during MBE strongly depends on the growth stoichiometry. The dislocation densities in the topmost layers matches the one of the overgrown substrate, which confirms the high quality MOVPE and MBE growth. Intentional and unintentional doping concentrations were investigated by secondary ion mass spectrometry (SIMS), electrical measurements on suitable test structures and the MOSFET itself. By this, the anticipated n-type doping of the layers by Si is well confirmed. Special attention is paid to the Mg doping of the p-GaN body layer, which is a complex topic by itself. Hydrogen passivation of magnesium plays an essential role, since only the active (hydrogen-free) Mg concentration determines the threshold voltage of the MOSFET and the blocking capability of the body diode. However, non-ohmic behaviour of the p-GaN contacts hinders the accurate analysis on transfer length method (TLM) as well as planar metal insulator semiconductor (MIS) capacitor structures. Instead the absolute and body bias dependent threshold voltage is investigated to get insight on the active Mg concentration. Trapping effects can lead to major uncertainties in the analysis of the absolute threshold voltage. The measurement of the body-bias dependency of the threshold voltage is used as advanced method to circumvent these limitations.

The gate module as central part of the fabrication process is patterned by a combination of chlorine-based dry etching and successive tetramethylammonium hydroxide (TMAH) wet etching. By tuning the process parameters of the dry etch critical pit formation at the trench bottom can be mitigated. The wet etch step further decorates those defects and shows a dependence of the vertical sidewall shape on the orientation with respect to a-plane or m-plane. The later causes a difference in the ON-state performance of the MOS-FETs with trenches aligned to the respective in-plane crystal directions. As gate dielectric atomic layer deposition (ALD) Al_2O_3 is used. Low gate leakage currents are obtained, due to the large band gap and convenient band alignment of the material. The quality of the gate dielectric/p-GaN channel interface with intended low trap densities strongly influences the threshold voltage and its stability.

The pseudo-vertical layer stack has a thin drift layer and much higher threading dislocation density, which degrades the reverse leakage and breakdown characteristics in the respective structures. Therefore, the OFF-state behaviour is preferentially focused on true-vertical structures with a drift layer thickness of 4 µm. The maximum electric field, which was achieved in the pn⁻-junction is estimated to be around 2.1 MV/cm, which is still 30 % below the reported material limit. The MOSFET breakdown voltage of 230 V is lower compared to the values of around 400 V obtained on the diode test structures, which indicates the lateral termination of the MOSFET as potential weak point. From double-sweep transfer measurements with relatively small hysteresis, steep subthreshold slope and a threshold voltage of 3 - 4 V a reasonably good Al_2O_3/GaN interface quality is indicated. In the conductive state a channel mobility of around 80 - 100 cm^2/Vs is estimated. The obtained value is comparable to device with additional overgrowth of the channel. Further enhancement of the OFF-state and ON-state characteristics is expected for optimization of the device termination and the high-k/GaN interface of the vertical trench gate, respectively. From the obtained results and dependencies key figures of an area efficient and competitive device design with thick drift layer is extrapolated. Finally, an outlook is given and advancement possibilities as well as technological limits are discussed.

Kurzfassung

Effiziente Energieumwandlung spielt eine wichtige Rolle um dem ständig steigenden Energieverbrauch unserer Gesellschaft zu begegnen. GaN basierte vertikale Leistungsfeldeffekttransistoren bieten exzellente Kenndaten für Leistungsschalter, auf Grund ihres Vermögens hohe Spannungen und Stromdichten bei gleichzeitig kleiner Flächennutzung zu verarbeiten. Diese Arbeit fokussiert sich auf das Konzept eines vertikalen Grabengate Metall-Oxid-Halbleiter Feldeffekttransistor (MOSFET) mit konzeptionellen Vorteilen in einer Bauelementherstellung vorgelagerten Epitaxie und Anreicherungscharakteristik. Der funktionsbestimmende Schichtstapel beinhaltet von unten eine n^+/n^--Drift/p-Body/n^+-Source GaN Schichtsequenz. Die Schichten bis zum p-GaN wurden mit metallorganischer Gasphasenepitaxie (MOVPE) gewachsen, wobei – spezifisch für diese Arbeit – die oberste n^+ Schicht mittels Molekularstrahlepitaxie (MBE) nach einem Temperschritt zur Wasserstoffausdiffusion gewachsen wurde. Dieser letzte Wachstumsschritt in einer wasserstofffreien Atmosphäre verhindert die erneute Absättigung von Mg-Dotanten durch Wasserstoff im p-GaN.
Herstellungsspezifische Herrausforderungen des Konzepts lassen sich der komplexen Integration, dem Aufbau von ohmschen Kontakten zu den funktionellen Lagen, der genauen Implementierung und Prozessierungsabfolge des Grabengatemoduls und dem lateralen Seitenabschluss des MOSFET zuordnen. Zunächst wurden als kosteneffiziente Testvehikel pseudovertikale Bauelemente basierend auf Saphirsubstraten mit schneller Rückkopplung zwischen Prozessierung und Wachstumsentwicklung genutzt. Im zweiten Schritt wurden volumenkristallbasierte echtvertikale Bauelemente auf GaN-Substraten mit dicker Driftlage hergestellt. Kooperative Teststrukturen erlauben die mikrostrukturelle und elektrische Charakterisierung der unterschiedlichen Bauelementeteile und stellen dadurch grundlegende Informationen und ein besseres Verständnis zur Funktionalität des kompletten MOSFET bereit. Wichtige Teststrukturen beinhalten beispielsweise den vertikalen $TiN/Al_2O_3/GaN$ Gatestapel oder die pn^- Bodydiode.

Da die Materialqualität großen Einfluss auf die letztendliche Baulementeleistungsfähigkeit hat, wird die Qualität des Wachstums und Überwachsens der aktiven Schichten mittels Abtasten der Oberflächenmorphologie durch Rasterkraftmikroskopie (AFM) untersucht. Endpunkte von Fadenversetzungen werden and der Oberfläche durch Defekte dekoriert, welche eine Abschätzung der Versetzungsdichte ermöglichen. Es wurde herausgefunden, dass der Dekorationstyp als Vertiefung oder Erhebung, während der MBE stark von der Wachstumsstöchiometrie bestimmt wird. Die Versetzungsdichte in den oberen Schichten entspricht dabei der des überwachsenen Substrats, was die hohe Qualität des MOVPE- und MBE-Wachstums bestätigt. Die bewussten und ungewünschten Dotierstoffkonzentrationen wurden mit Sekundärionen-Massenspektrometrie (SIMS), elektrischen Messungen an geeigneten Teststrukturen und am MOSFET selbst untersucht. Die angestrebte n-Typ Dotierungen der Schichten durch Si ist dadurch gut bestätigt. Spezielle Aufmerksamkeit kommt der Mg-Dotierung der p-GaN Bodyschicht zu, welche selbst eine komplexe Thematik darstellt. Wasserstoffabsättigung von Magnesium spielt eine entscheidende Rolle, da nur die aktive (wasserstofffreie) Mg-Konzentration die Einsatzspannung des MOSFET und die Sperrfähigkeit der Bodydiode bestimmt. Jedoch, behindert nichtohmsches Verhalten der p-GaN Kontakte eine akkurate Analyse an Transferlängen- (TLM) und planaren Metall-Isolator-Halbleiter (MIS)-Strukturen. Stattdessen wird die absolute und bodyspannungsabhängige Einsatzspannung untersucht, um Einblick in die aktive Mg-Konzentration zu bekommen. Ladungsträgereinfang kann zu dominierenden Unsicherheiten in der Analyse der absoluten Einsatzspannung führen, kann aber mit der Bodyspannungsmethode umgangen werden.

Das Gatemodul wird als zentraler Teil des Herstellungsprozesses mit einer chlorbasierten trockenchemischen Ätzung und nachfolgender nasschemischer Tetramethylammoniumhydroxid (TMAH) Ätzung hergestellt. Durch das optimieren der Prozessparameter der Trockenätzung kann die kritische Ausbildung von Vertiefungen am Grabenboden verhindert werden. Die nasschemische Ätzung dekoriert diese Defekte weiter und zeigt eine Gestaltabhängigkeit von der Ausrichtung der Seitenwand zur a- oder m-Ebene. Das Letztere führt zu Unterschieden in der Leistungsfähigkeit des MOSFET im AN-Zustand mit unterschiedlicher Ausrichtung der Gräben an den entsprechenden Kristallebenen. Als Gatedielektrikum wurde Al_2O_3 mittels Atomlagenabscheidung (ALD) hergestellt. Niedrige Gateleckströme werden durch eine große Bandlücke und günstige Bandausrichtung des Materials erreicht. Die Qualität der Gatedielektrikum/p-GaN-Kanalgrenzfläche mit angestrebter niedriger Störstellendichte beeinflusst stark die Einsatzspannung und deren Stabilität.

Der pseudovertikale Schichtstapel besitzt eine dünne Driftlage und wesentlich höhere Versetzungsdichte, was das Sperrverhalten und die Durchbruchscharakteristik der Strukturen verringert. Deshalb wird gezielt das Verhalten im AUS-Zustand an echtvertikalen Strukturen mit einer Driftlagendicke von 4 µm betrachtet. Die maximale elektrische Feldstärke, welche am pn$^-$-Übergang erreicht wurde, wird mit 2.1 MV/cm abgeschätzt, was nach wie vor 30 % unter dem berichteten Materiallimit liegt. Die MOSFET Durchbruchspannung von 230 V ist niedriger verglichen mit Werten um 400 V, ermittelt an Diodenteststrukturen. Dies deutet auf den lateralen Seitenabschluss des MOSFET als möglichen Schwachpunkt. Vorwärts- und Rückwärtsmessungen der Tranferkennlinie mit relativ kleiner Hysterese, steilem Unterschwellenanstieg und einer Einsatzspannung von 3 - 4 V lassen auf eine geeignete Al_2O_3/GaN-Grenzflächenqualität schließen. Im leitfähigen Zustand wird eine Ladungsträgerbeweglichkeit von 80 - 100 cm^2/Vs abgeschätzt, welche hoch für das hier gezeigte Baulement ohne zusätzlichen Kanalüberwachsschritt ist. Weitere Verbesserung der AUS- als auch AN-Charakteristik wird für eine Optimierung des Seitenabschlusses und der high-k/GaN-Grenzfläche der vertikalen Seitenwand erwartet. Aus den erhaltetenen Resultaten und Abhängigkeiten können Hauptkennwerte für ein flächeneffizientes und wettbewerbsfähiges Baulementedesign mit dicker Driftlage extrapoliert werden. Am Ende der Arbeit werden ein Ausblick gegeben und Verbesserungsmöglichkeiten als auch technologische Grenzen diskutiert.

Preamble

Die III/N Halbleiter bieten auf Grund ihrer vielfälltig erzeugbaren Kompositionen die Möglich- keit unterschiedlichste Eigenschaften auszubilden und erlauben damit eine umfangreiche experimentelle Grundlage, die gleichzeitig die Gebiete Leistungs-, Opto- und Hochfrequenzelektronik verbindet. Mit diesem Portfolio können unübertroffen zahlreiche Anwendungen erschlossen und kombiniert werden.

Es wäre schön, wenn man den Zutritt zu dem Gebiet der Halbleiter mit großer Bandlücke in einem Graphen erlangen könnte so wie es häufig durch die Darstellung vom spezifischen Flächenwiderstand über die Durchbruchspannung suggeriert wird. Leider ist die Thematik vielschichtiger und erfordert Erfahrung, Hintergrundwissen und ein Gespühr für die technologischen Möglichkeiten die am Ende entscheiden, ob ein Bauelement realisierbar und darüber hinaus wirtschaftlich herstellbar ist, oder nicht.

Die Intention die ich mit dieser Arbeit verfolge, ist nicht ein Material als überlegen darzustellen. Vielmehr ist es der Gedanke genau die oben genannten Fähigkeiten zu erlangen, zukünftig in meine Arbeit einfließen zu lassen und einen kleinen Beitrag in der effizienteren Nutzung von elektrischer Energie einem stetig steigenden Konsum entgegenzustellen. Die Gegenüberstellung verschiedener Materialien untereinander als auch mit etablierter Technologie hilft dabei die Möglichkeiten und Einsatzbereiche abzuschätzen, die von den zu erforschenden Materialsystemen ausgehen.

Nach wie vor bringen das Substratwachstum und die Epitaxie starke Limitationen für GaN Bauelemente mit sich. Kritisch, mit speziellem Blick auf vertikale Leistungsbauelemente, ist diesbezüglich die Entwicklung von wirtschaftlichen Methoden um qualitativ hochwertige GaN Substrate zu produzieren. Um die technologische Verbreitung und Auschöpfung der hervorrangenden Materialeigenschaften zu fördern, sollte deshalb stets ein rückgekoppelter Entwicklungsweg zwischen Bauelementetechnologie und Wachstum angestrebt werden.

Weiterhin finden bis jetzt Bauelementansätze basierend auf p-GaN Substraten keine starke Erwähnung in der Literatur. Die Entwicklung von intelligenten GaN-Leistungsbauelementen, die IGBT Technologie mit Niederspannungslogik basierend auf GaN oder Hybrid-Schaltkreisen kombiniert, erscheint aber nicht zuletzt in Verbindung mit der Hochspannungsenergieübertragung erstrebenswert und motivierend.

Table of contents

1 Motivation and boundary conditions

The continuously increasing consumption of electric energy is one of the major issues of this century. A large amount of the demand is still covered by non-renewable sources and a high amount of power is wasted in conversion to suitable voltage, current and frequency levels, used in various applications. Besides the need for sustainable power generation the development of efficient and robust systems for the distribution of energy is needed.

Already 30 years ago a broad range of materials with favourable properties for power electronics such as semiconductors with a wide band gap were discovered and heavily investigated. Partially some candidates are developed to products already spreading into application. The expectations on newly introduced materials are not limited to their electric performance, but also thermal and geometric aspects are interesting for electric devices with high power density. The GaN based technology benefits from several advantages accommodated with the inherent material properties of gallium nitride (GaN) and its compounds with aluminium nitride (AlN) and indium nitride (InN), e.g. the high channel mobility of the two dimensional electron gas (2DEG).

In the following subsections an overview of competing material systems and device concepts will be given. On one hand, these materials belong to strongly developed branches such as silicon (Si) or gallium arsenide (GaAs) technology. On the other hand, rather new candidates such as silicon carbide (SiC), gallium nitride and diamond (C) offer great potential. Although, these materials show superior properties the usage beyond niche applications depends strongly on the availability of substrate material and epitaxy (e.g. substrate size, substrate type, doping), material quality (e.g. wafer bow, surface morphology, threading dislocation density (TDD)) and on its processing efforts (e.g. mechanical processing, interface engineering, contact formation, thermal treatment etc.) and thus on the cost of the complete product line. Beside these technological aspects, the cost for fabrication of devices can be drastically reduced if the components are compatible to already established Si complementary metal-oxide-semiconductor (CMOS) technology and thus manufacturing sites can be shared.

1.1 A comparison of competitive semiconductor materials

Several years ago the two material systems SiC and GaN were started to be studied on their potential for electronic applications. The superior material properties, in particular in high power applications compared to Si and GaAs, have already lead to a partial successful integration in commercially available applications. More recently diamond was started to be investigated, due to its superior thermal and electrical material properties such as negative electron affinity [1] and very high breakdown field strength and thermal conductivity. However, the demanding growth of larger substrates with appropriate doping of both, n- and p-type layers as well as challenges in the formation of ohmic contacts seem to impede a widespread application in the short to mid time frame.

The appearance of SiC and its polytypes is manifold. One of the most used representatives in electronic applications is hexagonal SiC with the stacking sequence ABCB (4H-SiC). This material derivative has quite similar band gap, a slightly lower mobility and electric breakdown field compared to GaN. Hence, it is a direct competitor. Another wide band gap semiconductor is zinc oxide (ZnO), which has similar carrier mobilities and band gap properties compared to GaN. It is well known from thin-film technology for transparent devices, sensor applications or acoustic wave devices, but is also proposed for high-power electronic applications. Difficulties for the ZnO technology arise from p-type doping, similarly as for GaN. Additionally the thermal conductivity in comparatively low. Further information is can be found elsewhere [2, 3].

Although, GaN is used in RF-devices with a voltage class below < 200 V since years, the more mature SiC growth technology and the available sizes and qualities (TDD < 10 cm^{-2}) of SiC substrates have already lead to stronger use of SiC in power electronics for energy conversion applications. Typical examples are Schottky diodes, junction field effect transistors (JFETs) and MOSFETs. However, difficulties still arise from p-type doping, epitaxy, contact formation as well as handling of the high thermal budget necessary during the epitaxy and after implantation.

Tab. 1.1: Physical parameters, crystal structure and dopant properties of C, SiC, GaN, GaAs and Si at 300 K [4, 5, 6, 7, 8, 9]. The crystal structures relate to the mostly used material modification such is 4H-SiC, one of the most prominent polytypes next to 6H- or 3C- in the SiC material family. As an overview the shallowest dopants are given, with typical activation energies for high doping levels. However, these values are only for general discussion and can depend strongly on the growth method and doping concentration. The substrate sizes denote the largest typically available commercial substrate diameter.

Property/Material	C	4H-SiC	GaN	GaAs	Si
Band type	indirect	indirect	direct	direct	indirect
Crystal structure	diamond	wurtzite	wurtzite	zinc blende	diamond
E_g [eV]	5.45	3.26	3.44	1.43	1.12
$\mu_{n/p}$ [cm^2/Vs]	1800/1200	800/115	1400/80	8500/400	1450/450
E_{Br} [MV/cm]	10	2.5	3.3	0.6	0.3
v_{sat} [$\cdot 10^7$ cm/s]	2.2	2	2.3	1.2	1
χ [eV]	0.2	3.5	4.1	4.1	4.1
ϵ_r	5.8	9.7	9	13	11.7
κ [W/mK]	2400	370	160	50	150
Shallowest donor	P	N	Si	Si	P
$E_C - E_D$ [meV]	100	45	25	6	45
Shallowest acceptor	B	Al	Mg	Zn, C	B
$E_A - E_V$ [meV]	370	200	150	20	45
Subs. size d_{Sub}	1 in.	200 mm	100 mm	200 mm	450 mm

A comprehension of the material properties as well as common substrate sizes and typically observed activation energies of the shallowest dopants are given in Tab. 1.1. The symbol E_g is the band gap energy, $\mu_{n/p}$ represents the mobility for electrons and holes in the bulk material, E_{Br} is the breakdown electric field, v_{sat} the saturation velocity, χ the electron affinity, ϵ_r the relative permittivity, κ the heat conductivity, E_C the conduction band energy, E_D the donor energy, E_A the acceptor energy, E_V the valence band energy, and d_{sub} the substrate diameter. Overall the comparison of GaN to Si and GaAs suffers from the advantage of decades of investigations and development for the later two materials. Most of the industrial equipment was/is developed for Si-based devices with large substrate diameter and new tools suitable to supply the necessary processing features, e.g. high temperature annealing for implant post treatment and rapid thermal processing evolve slowly. However, stronger development and more common use of alternative materials support this development.

Especially silicon as single element semiconductor offers great possibilities in metal contact formation with Schottky junctions and ohmic contacts, due to silicide formation with various metals. Fruther, its native oxide adds convenient interface engineering possibilities. These – almost taken for granted – features reveal to be one of the most critical steps in new device technologies. In particular, due to the occurrence of high energy barriers associated with p-type doping in wide band gap material. Strong differences also occur from the difficulties in doping by implantation. High temperature annealing is needed to activate implanted doping agents and to restore the crystal integrity. Nevertheless, a strong crystal degradation is typically observed and partially remains. Arbitrary doping profiles are thus not easy to achieve. Complex doped well structures and doping gradients are yet impractical. Instead, intended doping profiles are usually set up during the epitaxial growth of the functional layers. This limits the design freedom in lateral as well as vertical direction, but provides better crystal quality. Other approaches like epitaxial lateral or selective overgrowth add additional complexity.

In the next section an overview over different device concepts for vertical GaN power devices will be given. Usually a set of key device parameters is important in the specified operation regime or application. For power-MOSFETs major preferential electrical parameters are:

- a high device blocking voltage (V_{BR})
- a low ON-resistance (R_{DSon}) defining the ON-state loss and current capability
- normally-OFF operation ($V_{th} > 0$ V)
- fast switching with instant transitions, i.e. high du/dt and di/dt capability
- high temperature stability
- long term reliability

Although this work focuses on static properties of (pseudo-)vertical MOSFETs other device concepts have been developed based on these thoughts.

1.2 Vertical GaN device concepts

In the previous section an overview about different power semiconductors has been given. Beside classical power switches, of course different other device classes can be addressed by wide band gap semiconductors, such as Schottky and pn-diodes. Though, in this work data about similar devices is discussed, they are mainly used as characterisation structures. Once a certain semiconductor material is focused several different switching device concepts are available, which shall be explained closer in the following. Classically Si-power FETs are classified by their gating principle (junction or MIS), their channel mode (enhancement or depletion) and their channel polarity (n- or p-type). The variety of device types based on GaN is subdivided in case of the insertion of heterostructures, but can be structured in the following way with a successive classification of the device features.

- Principle guidance of the current with respect to the substrate i.e. lateral or vertical
- Gate orientation related to the surface of the substrate i.e. lateral, slanted or vertical
- Gating type i.e. junction or insulated
- Channel type i.e. accumulation or inversion
- Channel polarity i.e. n-type or p-type
- Implementation of the gate i.e. directly or using a heterostructure

The here presented vertical MOSFET is thus a vertical device with vertical insulated gate formed with an n-channel inversion layer in vicinity of a dielectric/p-GaN interface. Many other configurations are imaginable but not all combinations within this scheme are feasible, e.g. a junction gate in combination with an inversion channel. However, some concepts have asserted themselves and shall be outlined in the following.

Fig. 1.1 gives an overview of proposed basic devices concepts [10][11]. Modifications in order to improve their performance or to change details in the device behaviour are mostly linked to the threshold voltage [12, 13], interface quality [14] or device insulation [15]. They are explained alongside.

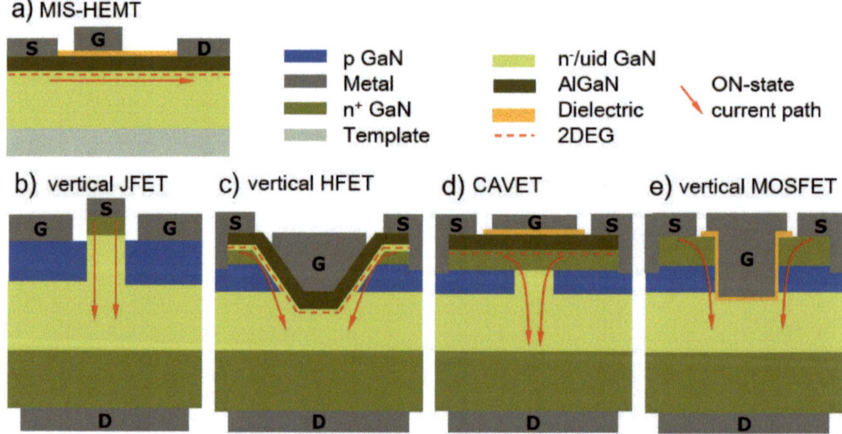

Fig. 1.1: Schematic cross sections of different GaN power device concepts. Topmost, in a) the lateral metal insulator semiconductor high electron mobility (MIS-HEMT) is shown. Below, the four different concepts b) vertical JFET, c) vertical heterostructure FET (VHFET), d) current aperture vertical transistor (CAVET) and e) vertical MOSFET are presented in their basic design. The letters G, S and D represent the gate, source and drain metal contacts, respectively.

While the vertical MOSFET shows intrinsically normally-OFF behaviour caused by the formation of an inversion channel within the p-GaN body layer in ON-state, the JFET, VHFET and CAVET are not necessarily normally-OFF devices. Their threshold voltage strongly depends on the device geometry, gate electrode and doping concentrations. But beside geometric optimization the gate contact work function is very often modified to increase the threshold voltage of the devices. A higher gate electrode work function can be achieved by either a high work function gate metal or utilizing a p-GaN gate electrode. With the later an even higher effective work function can be achieved. This enables more design flexibility, but induces another growth step for this functional layer. Further a compromise between gate leakage current and gate coupling is necessary, when it is combined with a replacement or thinning of the gate dielectric. In order to stronger discriminate between potential applications the different concepts are separately discussed according to Fig. 1.2.

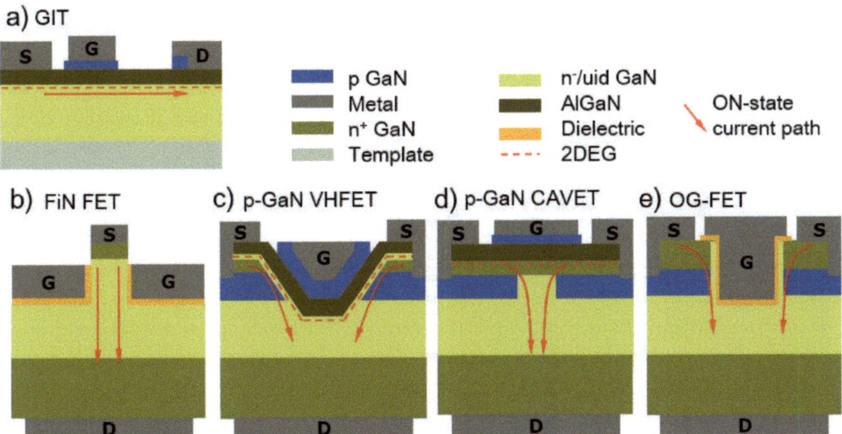

Fig. 1.2: Schematic cross sections of the different modified GaN power device concepts: a) GIT, b) FiN FET, c) p-GaN VHFET, d) p-GaN CAVET and e) OG-FET. Modifications are done in order to lower leakage currents and/or increase the threshold and breakdown voltage as well as the channel mobility of the devices.

- For completeness an example for a lateral device, is presented with the **MIS-HEMT** [16, 17, 18]. Utilizing an insulated gate to avoid the leakage associated with the Schottky gate of a HEMT without gate dielectric [19], these high mobility devices provide normally-ON operation, not favourable for fail-safe operation in high power applications. Variations of the structure with gate recess [20], gate implant [21] or a p-GaN gate contact [22, 23] have been reported in order to achieve normally-OFF operation. As example the gate injection transistor (GIT) with p-GaN gate stack is shown in Fig.1.2. Similar strategies can be applied for the shown vertical device concepts with an AlGaN/GaN heterostructure.

Due to the high electron mobility in the 2DEG within the unintentionally doped (UID) channel region the (MIS)-HEMT is an excellent candidate for high frequency, medium power operation. Due to the lateral device orientation and thus the scaling of the device length with the targeted device breakdown voltage, the concept is most likely limited to device classes of around 1000 V.

The vicinity of the channel to the surface deteriorates the thermal conduction of dissipated heat from the channel and introduces surficial effects like trapping. Phenomena often reported as current collaps, knee walkout or gate and drain lag [24, 25, 26] result. Strong electric stress at the gate edge can be supressed by the application of field plates [27, 28]. This, in turn comes along with increased gate-source and gate-drain capacitances [29], though these parasitics are still much lower compared to silicon devices. Due to the significant cost reduction by growth of the needed AlGaN/GaN layer stack on foreign substrates such as silicon wafers successful development was conducted in the last years. A new GaN-HEMT based device class for voltages up to 600 V is already established.

- A finned vertical accumulation channel is utilized within the **vertical JFET** [30]. Along with the need for high resolution lithography or self aligning processes a positive threshold voltage can be achieved by inherent depletion in lateral direction into the submicron broad channel. To exclude tail currents affiliated with the junction reverse recovery behavior and to substantially reduce leakage currents the p-GaN can be replaced by a MIS-structure similar to the vertical MOSFET gate [15, 31], see the Fin FET in Fig. 1.2. In this case a three terminal device with high bulk-channel mobility is created with bidirectional blocking capability and reduced complexity in the GaN growth scheme. The small area for the thermal attachment of the fin structure to the substrate and potential electric field peaks at the fin base counteract the lateral down-scaling potential of these structures.

- A slanted 2DEG channel is used in the **vertical HFET** intended to be of normally-OFF type. However, a good compromise between channel mobility and p-GaN doping concentration as well as concentration, slope and thickness of the AlGaN barrier is essential [32, 33]. Similar to the lateral GIT the threshold voltage of this device can be increased by adding a p-GaN gate layer as shown for the p-GaN VHFET in Fig. 1.2 [15]. Compared to the other vertical concepts a scaling of the lateral device dimensions with changing device breakdown voltage is obvious. Reasons for this can be found in the slanted gate structure and the necessity to increase the thickness of the body layer for higher device breakdown voltage ratings. Hence, the scalability of the device is limited, especially for thicker p-GaN layers in very high voltage designs. For a lower voltage range the low channel resistance and high mobility can potentially counterbalance the demanding growth process of the slanted heterostructure.

- The **current aperture vertical transistor** utilizes a source sided 2DEG and low doped channel confined by a current blocking aperture (here p-GaN) leading to a high channel mobility, good thermal managment but concurrently normally-ON characteristics and demanding device growth and processing [34, 35]. The gate/blocking layer overlap constrains the down-scaling capability in lateral direction. This drawback can be less important for application of this device concept for very high voltages and power densities. Also here, a p-GaN gate increases the effective gate work function to shift the threshold voltage towards higher positive values.

- Proleptic to chapters 2 and 3 a short comprehension of the **vertical MOSFET** shall be given here. Though good scaleablity, normally-OFF operation and good thermal coupling of the channel to the bulk material is promoted, a low channel mobility and a demanding gate stack processing has to be taken into account for this device concept. In order to improve the channel mobility and trapping characteristics an additional UID GaN-layer can be grown on the surface of the vertical channel. Such a device with GaN interlayer and gate oxide was shown in [36] and refered to as OG-FET. One drawback for this approach is the increased process complexity and the reduction of the threshold voltage.

One common problem for all device types in GaN is still the availability, cost, size and quality of the substrate material. Fig. 1.3 presents the cost and usual sizes of commercially available substrates made of various epitaxial materials. Up to now bulk GaN substrates show the best crystal quality in terms of TDD. But they are rarely used for device process development due to their high cost for reasonable sizes of ≥ 2 inch. Rather thin heteroepitaxial layers on foreign substrate material such as sapphire, silicon carbide or in particular silicon can reduce the substrate cost drastically, but come along with a high TDD and mechanical stress inside the layers which has to be managed by special buffer structures, as reference see [37]. In lateral device concepts these structure are additionally used to improve the vertical device insulation. Fortunately the 2DEG mobility at room temperature is typically limited by phonon scattering and thus not very sensitive to TDD in a reasonable range.

For vertical devices the application of foreign substrate based GaN templates is more delicate and generally high voltage designs are not possible, due to the limited layer thickness and high TDD. However, they are suitable candidates to set up a process technology and investigate different building blocks. Hence, these devices are used in this work e.g. to evaluate the gate module.

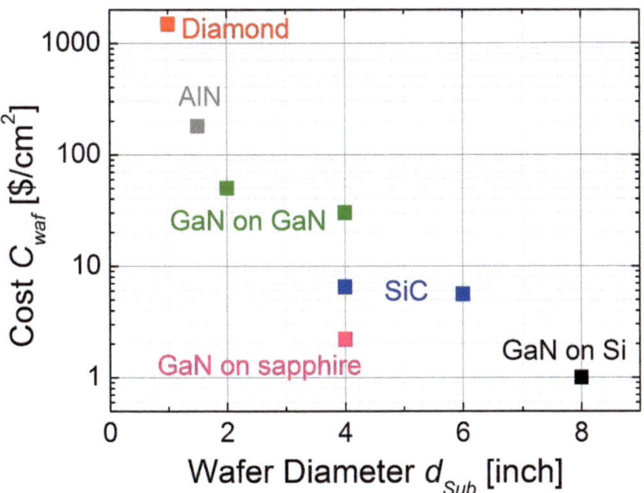

Fig. 1.3: Cost per square centimeter including epitaxial and wafer cost versus size of the commercially available substrates. The data is reprinted from [38].

1.3 Target application for power switches

In this section an introduction to the application of power switches shall be given. Loss mechanisms are outlined and ideal characteristics of the devices are described. The most prominent target application for the here presented vertical GaN trench MOSFETs is the usage as semiconductor switches in a mid voltage and high power environment. For example the usage in energy conversion circuits in the voltage range from 600 V up to 5 kV is intended. In this range currently silicon-based insulated gate bipolar transistor (IGBTs) are dominant, due to their optimized trade-off between ON-state loss and voltage blocking capability. The bipolar collector-emitter junctions results in a collector-emitter saturation voltage V_{CEsat} and not as for unipolar MOSFETs in a classical ON-resistance. Similar to pure bipolar transistors this voltage is typical for a certain semiconductor material and doping profile. It leads to a minimum constant voltage drop and changes only slightly over a wide range of operation currents (I) and different blocking voltage specifications. Thus, the static ON-losses scale linearly with $V_{CEsat} \cdot I$. The MOSFETs R_{DSon} instead scales with the blocking voltage and the ON-losses increase with $R_{DSon} \cdot I^2$, which make the IGBT concept advantageous above around 400 - 800 V in Si technology.

The drawback coming along with that is the lower switching speed of IGBTs, due to the bipolar nature causing a high base and junction recovery current (tail current). The application of Si IGBTs is limited to around 7 kV. At high operation voltages typically only thyristor based devices remain practical. By changing the semiconductor material to GaN or SiC in order to decrease the specific R_{DSon} per blocking voltage and area the trade-off transition point moves towards higher voltages and the MOSFET concept gains additional voltage margin. The lower specific R_{DSon} for unipolar GaN or SiC devices enables the competition to the Si IGBT with higher possible switching speed. With that, the size of peripheral passive components such as capacitors or inductors can be reduced and the system power density and efficiency increases.

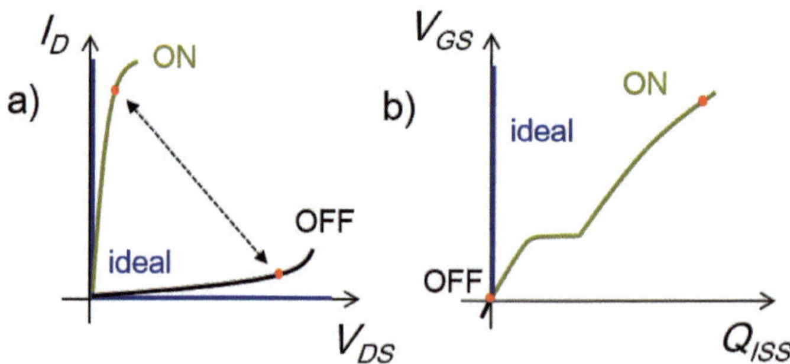

Fig. 1.4: Schematic output (a), $I_D(V_{DS})$) and input gate-charge (b), $V_{GS}(Q_{iss})$) characteristics of a power switch with its ideal (blue) and realistic (green and black) operating conditions, typical operating points (red) in application are shown for fully ON and OFF-state

Focusing now on the MOSFET, on the one side quasi static properties like R_{DSon}, threshold voltage V_{th} and breakdown voltage V_{Br} are important parameters, since they determine the static losses and the application case in a more general way. On the other side, dynamically acting parameters such as gate-source (C_{GS}) and gate-drain capacitance (C_{GD}) are crucial for the later device operation. They are rather important for performance and safety issues, e.g. switching loss and/or parasitic gate bouncing, due to reverse coupling between drain and gate or even re-turn on. This sets the maximum switching frequency and the duty cycle. Schematic characteristics are shown in Fig. 1.4 with typical operation points in ON- and OFF-state.

As previously introduced the losses during operation can be classified in static and dynamic. The static losses (P_S) are described with indexes "on" and "off" for the ON- and OFF-state losses, respectively, by

$$P_S = (R_{DSon}I_{Don}^2 + I_{Gon}V_{GSon})D_f + (I_{Goff}V_{GSoff} + I_{Doff}V_{DSoff})(1 - D_f). \tag{1.1}$$

The terms for the gate are mostly negligibly small, especially when no additional negative gate bias is applied in OFF-state. This is convenient to decrease the time needed to discharge the gate during switching. The most prominent contributions are the losses associated with the drain. In OFF-state losses arise from the high voltage applied between drain and source and the drain leakage current. The ON-state losses are produced due to the residual voltage dropping on the device R_{DSon} at high drain current. Since the device is usually switched between OFF- and ON-state it remains only a portion of the time in the respective state. A certain duty cycle $D_f = t_{on}/(t_{off} + t_{on})$ for the ON-time or $1 - D_f$ for the OFF-time has to be taken into account. The dynamic power loss in contrary depends on the switching frequency and speed and thus increases with faster and more frequent switching. It can be approximated by Equation (1.2) with the input charge (Q_{ISS}) and the output charge (Q_{OSS}) from Equation (1.3) and (1.4). This approximation assumes, that a switching event is fast compared to the duty cycle and completes at the full ON- or OFF-state. Intermediate gate-source voltages for longer times lead to extensive power dissipation in the switch and usually cause degradation or destruction. The dynamic loss can be calculated by:

$$P_D = f_s \left(Q_{ISS}\Delta V_{GS} + Q_{OSS}\Delta V_{DS} \right). \tag{1.2}$$

The capacitances depend on the voltages applied between the respective terminals. Hence, it is more convenient to directly use the charge e.g. obtained by gate-charge measurements [39]. It can be calculated by

$$Q_{ISS} = \int_V (C_{GS})dV_{GS} + \int_V (C_{GD})dV_{GD} \tag{1.3}$$

and

$$Q_{OSS} = \int_V (C_{DS})dV_{DS} + \int_V (C_{GD})dV_{GD}, \tag{1.4}$$

where C_{GS} and C_{GD} are a function of V_{GS} and V_{GD} respectively. From these equations it can be seen that both charges are coupled by C_{GD} and V_{GD}. If the device is switched between a high and a low V_{DS} state (V_{GD} equals approximately V_{DS} in this case) and the driving circuit for the gate cannot supply enough current to counterbalance Q_{GD} e.g. due to series resistance contributions, the charge is transferred into the gate and can lead to a strong V_{GS} swing, which counteracts the actual switching direction. Ringing or oszillation is created and causes undefined switching, which is extremely unfavorable for the switch and the periphery. The transformation of the gate-drain capacitance into the gate-source circuit is sometimes refered to as "Miller effect" with C_{GD} as the "Miller capacitance". It is stronger pronounced for faster voltage changes (dV_{DS}/dt) [40]. One possible way to avoid losses associated to C_{GD} is zero voltage switching. The idea behind is to let V_{DG} settle to ≤ 0 V before an OFF-switching event occurs, e.g. in resonant power converter topologies.

A further aspect concerning the fail safe circuit operation is the resistivity against external operation faults. An occassionally or accidently occuring operation point outside of the defined device specification e.g. a voltage spike causing a shortly increased loss should be captured by the switch without destruction. In case the device is damaged a safe quiescent default operation point is needed, which does not allow the destruction of further circuitry. In this respect, normally-ON devices suffer from low quiescent ON-resistances, if no gate-source voltage is applied. This results in a quasi-short between drain and source. Normally-OFF operation is beneficial, due to the inherent high resistive state of the switch during a gate-sided failure or when no gate-source voltage is applied. Another approach to face both, normally-ON behaviour and miller effect but with higher component count, is the use of a cascode circuit. Typically a GaN normally-ON HEMT is used in conjunction with a Si MOSFET for this circuitry [41]. The cascaded HEMT stabilizes the drain-source voltage of the low side Si MOSFET. Thus the term $\int_V (C_{GD})dV_{GD}$ in Equation (1.3) is very small for the Si MOSFET with positive threshold voltage. A schematic cascode arrangement is presented in Fig. 1.5.

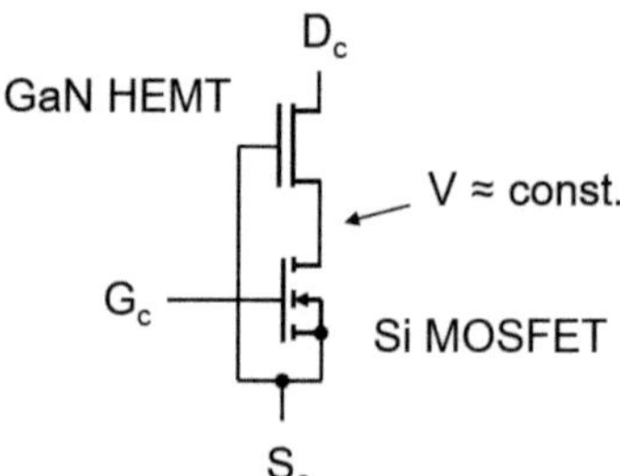

Fig. 1.5: Schematic circuitry of a cascode arrangement including a GaN HEMT in common-gate and a Si MOSFET in common-source circuitry. The resulting external cascode source, drain and gate terminals are denoted by S_c, D_c and G_c, respectively.

The compromise is a higher component count and a non-monolithic arrangement of a Si and GaN FET. Additionally the ON-resistance of the cascode results from the series connection of the GaN-HEMT and Si-MOSFET.

2 The vertical GaN MOSFET concept

In the previous chapter the boundaries and the placement of the III/N technology concerning different competing materials was presented. In this section detailed analysis of the device concept and analytical approximations for the expected device performance are given. The physical and mathematical calculation of device properties is limited by the set model and it's boundary conditions. The material quality, typically assumed to be ideal in theoretical estimations, is a hugh factor for deviations from the real to the idealized device behaviour. The interactions of various imperfections likewise dislocations, impurities, deep dopants with carrier transport or breakdown behaviour and thus static and dynamic device operation are manifold. Therefore physical modelling is mostly restricted to particular problems. Nevertheless, the approximation and optimization can be understood as performance limit and shows the potential of the material system. Further, properly modelled problems allow deeper investigations and promote a better understanding and experience.

In the following subsections different structural parts of the MOSFET are addressed in order to separately estimate the ON- and OFF-state performance. Granting access to systematic structure design depending on the application requirements shall be target in this context. The influence of deficiencies originating from material imperfections will be discussed alongside. Therefore a short introduction to the incomplete ionization of Magnesium, typically used for p-type doping is given in the next section.

2.1 Incomplete ionization of dopants

Compared to silicon technology, where incomplete ionization is mostly not significant, wide band gap semiconductor suffer from incomplete dopant ionization and/or dopant passivation for p-type doping. In order to explain the impact of these phenomena on the device properties an overview is given in this section.

Different requirements are neccessary to qualify a certain element as dopand in a semiconductor [42]:

- adequate solubility
- incorporation without passivation on the appropriate crystal lattice site within the semiconductor
- low activation energy (< 5 kT)

The Boltzmann constant and temperature is denoted by k and T, respectively. Within the silicon technology incomplete ionization is rarely reported to be problematic. The used dopands have generally low activation energies and show appropriate solubility and incorporation. As one exception indium is sometimes used, due to its low diffusivity. It is a p-type dopant with relatively high activation energy of around 160 meV [43, 44].

In the 1980s, one of the most demanding challenges for GaN devices, was the generation of p-type material. Hydrogen, which is incorporated during MOVPE passivates the acceptor states in complexes. With the discovery of an efficient way to destroy these complexes and remove the hydrogen it became possible to achieve hole conduction in the widely used

MOVPE layers. To that time, extensive interest in this topic arose from the development of light emitting structures [45, 46]. In principle Mg, Zn, C, Be, Ca or Cd can be used as acceptors in GaN, though by far the most important and commonly applied is Mg [47, 48, 49]. Its utilization and importance for this work reasons the particular focus on this kind of p-type dopant. During the vapour phase epitaxy of GaN the incorporation of Mg is typically associated with the presence of hydrogen inside the growth atmosphere caused by either molecular hydrogen as carrier gas and/or additionally hydrogen orginating from the precursor material. A weakly bound Mg-H complex is formed in this case within the GaN lattice, resulting in an initially semi-insulating (SI) material [50, 51, 52]. A low electron energy radiation or thermal treatment can be applied to break the complex. The high diffusivity of H in GaN and the application of H-lean conditions during this treatment allows the hydrogen to be removed first from the acceptors and moreover out of the material. This procedure creates a p-type GaN layer, due to the formation of Mg acceptor states with 150 meV - 240 meV activation energy (E_A) [53, 54, 55, 56, 57] as extracted from different characterisation methods. From theory Mg as group II material acts on both lattice sites as acceptor, but typically Mg is considered to be incorporated at the gallium lattice site [58, 59]. The process of hydrogen removal is often called 'activation', but should not be mixed up with the activation or ionization of Mg-acceptors.

To fully clarify the consequences and properties of the Mg-doped GaN layers the conceptional naming and classification is discussed in the following. The totally incorporated concentration of Mg during the growth is called N_{Mg0} and is not necessarily located on the intended lattice site (Ga-site), but can be located at the nitrogen lattice site Mg_N or can form clusters or interstitials. All of the latter are represented by N_{Mgi} in

$$N_{Mg0} = N_{Mgi} + N_{Mg-H} + N_A \approx N_{Mg-H} + N_A \qquad (2.1)$$

The partial amount of Mg located on the Ga-site can be passivated by hydrogen and is called N_{Mg-H}. The right-most sum in (2.1) represents the concentration contributions under a reasonable assumption of negligible N_{Mgi}, due to their low formation probability. Depending on the particular treatment to remove the hydrogen, the obtained unpassivated concentration of the acceptor concentration N_A is formed. Although, this concentration represents the amount of Mg, which can potentially accept an electron, only a small amount is ionized at 300 K in equilibrium. The activation efficiency A_A can be estimated by:

$$A_A = \frac{N_{A-}}{N_A} = \frac{1}{1 + g_h exp\left(\frac{E_A - E_F}{kT}\right)}, \qquad (2.2)$$

where E_F is the Fermi energy and g_h the degeneracy factor for holes. Caused by the high activation energy of the acceptor state A_A is the range of $1 - 3\%$ at room temperature. From the previous definition Mg is thus not an ideal dopant, but it is commonly used. It is just the best one available.

The ionization of acceptors is characterised by bounding an electron. All ionized acceptors are negatively charged and are represented by N_{A-}. They represent e.g. the major depletion charge of a space charge region. Sometimes, though not completely correct, this concentration is also referred to as activated Magnesium concentration. But it is rather meant to be the hydrogen free concentration. In equilibrium or flat band conditions the presence of N_A and its thermal activation causes a hole concentration p, which in absence of further impurities, equals N_{A-} in the bulk p-GaN. However, N_{A-} strongly depends on the value $E_A - E_F$ and thus differs in different structural parts of a device like space charge regions, regions with different background donor concentration or bulk p-GaN. Figure 2.1 sketches the difference of the acceptor activation in the bulk layer and in a space charge region under inversion of a MIS-structure.

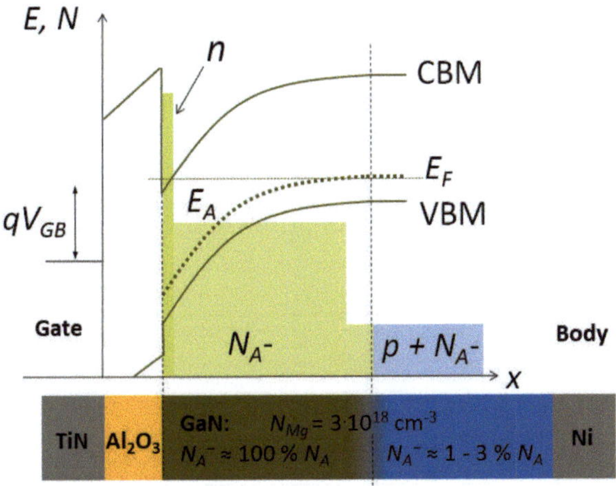

Fig. 2.1: Energy (E, CBM, VBM, E_F, E_A, qV_{GB}) and charge concentration (N, n, N_{A-}, p) versus depth (x) and schematic of the MIS-structure on p-GaN in strong inversion. The electrostatic charge induced by the applied positive gate-body voltage V_{BG} is formed by ionized acceptors in the volume of the space charge region and by electrons forming a sheet charge layer at the dielectric/GaN interface.

N_{Mg} can be determined by secondary ion mass spectrocopy (SIMS), whereas N_{A-} and p can be investigated by electrical characterisation. The concentration of N_A is not directly accessible, but can be estimated with the knowledge of N_{A-} from measurements utilizing a depletion layer and N_{Mg0}. Interstitial Mg and Mg$_N$ are commonly neglected, if low or medium Mg concentrations are used [60].

With an activation energy as described earlier, on the one side the ionized acceptor concentration depends strongly on the temperature in typical operating conditions, but also on the Fermi energy as shown in Equation (2.2). In bulk p-GaN the Fermi level can be numerically calculated with the charge neutrality equation for equilibrium conditions [42]:

$$N_{D+} + p = N_{A-} + n \qquad (2.3)$$

and its reformulation:

$$\frac{N_D}{1 + g_e exp\left(\frac{E_F - E_D}{kT}\right)} + N_V exp\left(\frac{E_V - E_F}{kT}\right) = \frac{N_A}{1 + g_h exp\left(\frac{E_A - E_F}{kT}\right)} + N_C exp\left(\frac{E_F - E_C}{kT}\right). \qquad (2.4)$$

By subsequent calculation of the second term (for p) in Equation (2.4) the hole concentration can be obtained. N_D and N_{D+} represent (according to N_A and N_{A-}) the donor and ionized donor concentration, respectively. The effective density of states in the conduction and valence band are taken by $N_C = 4.3 \cdot 10^{15} \, T^{3/2}$ cm^{-3} and $N_V = 8.9 \cdot 10^{15}$ $T^{3/2}$ cm^{-3}, respectively. The semiconductor degeneracy factor for holes and electrons are represented by $g_h = 4$ and $g_e = 2$. For space charge regions this equation is simplified, due to negligible carrier concentrations and/or full ionization of the doping agents. For example, to estimate the threshold voltage of a MOSFET solely N_{A-} has to be taken into account.

2.2 The pseudo-vertical approach

Before specific focus is put on detailed considerations for the device OFF- and ON-state generally applicable to vertical devices the pseudo-vertical device approach shall be introduced. On the first view the principles behind the design differences of the true-vertical and pseudo-vertical MOSFET are not very strong, but compromises have to be made from the growth and processing point of view. Practically, a pseudo-vertical device cannot outperform a similar designed vertical device. It is rather a very suitable test vehicle in order to proof the process feasibility and investigate single device features, like the ohmic contact formation, the dielectric/GaN interface or the properties of different functional layers. Feedback for the epitaxial growth and substrate preparation can be gained. Additionally it combines considerations about the availability of substrates and their cost as well as changes in processing and device characteristics with the technological development. The need for rare and costly free-standing bulk substrates for a true-vertical device can be seen as second step in order to proof the results from the pseudo-vertical devices.

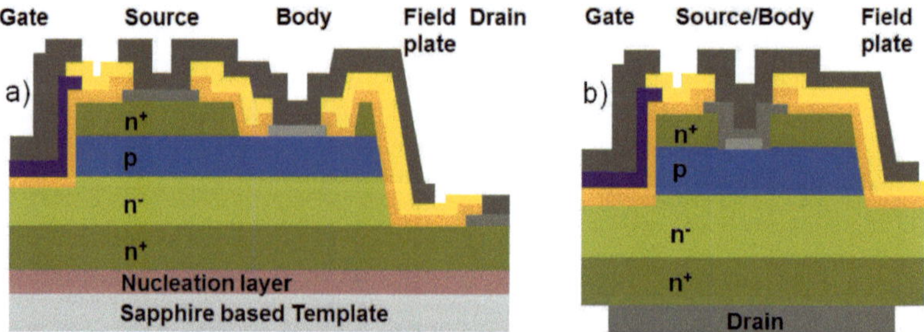

Fig. 2.2: Schematic cross sectional view of a symmetrical half sided pseudo-vertical MOSFET (a)) as it is used in this work as characterisation structure and true-vertical MOSFET (b)) as it would be considered for the final application.

The major changes and compromises for the pseudo-vertical device design shall be outlined in the following:

- omitting of a back-side metal deposition, due to a top-sided drain contact
- deep etching to the bottom n$^+$ drain layer to fabricate a drain contact
- a rather thin drift layer to maintain the ability to access the n$^+$ drain layer

- an increased drain layer resistance, due to the lateral positioning of the drain contact
- typically higher dislocation density on the heteroepitaxial VPE GaN templates
- potential increase of leakage currents along dislocations
- a much lower breakdown voltage compared to true-vertical devices, due to the thinner drift layer and a higher defect density

As described, investigations on the contacts to the p-GaN body layer, the gate module and characterisation of the differently doped GaN layers can be used to estimate the device behaviour of a true-vertical MOSFET. Feedback to the growth procedure can be obtained in order to improve the layer properties such as doping, impurity and dislocation density. A schematic cross sectional view of the two different device types is presented in Fig. 2.2. For convenience in b) the true-vertical device is shown as it is targeted for later power device processing. To minimize the used active device area it includes a field plate and a body contact merged with the source. The pseudo-vertical device (a), in contrary, is shown as a characterisation structure with a separate body contact. The drain contact is placed at the top side of the device, as described previously. A field plate can be utilized, however, its application is more crucial in true-vertical devices with much higher target operation voltages and larger electric fields.

In general, the best achievable OFF-state properties are always in conflict with the ON-state characteristics of the device. A thick drift layer with low doping concentration can buffer higher voltages, but simultaneously it increases the R_{DSon} of the device by its series resistance. A compromise depending on the application is needed.

2.3 Considerations for the device OFF-state

In this section considerations concerning the OFF-state of the device are discussed. Nevertheless, these aspects are in close conjunction with the device ON-state described in Section 2.4. Fig. 2.3 summarizes the two main regions, which are interesting for the OFF-state design, the pn⁻ junction area (a and b) and the MIS-structure related to the gate trench and field plate area (c). Referring to this figure, both regions will separately be discussed in the next sections. The fabrication of separated test structures like pn⁻ diode and MIS-structures is described in Section 3.4.3. These separated structures are used for characterisation of central functional device parts independently to the entire MOSFET. Their specific influence can thus be examined easier. Fig. 2.3 gives an overview of the device-internal active regions under OFF-state conditions.Alongside to this figure in the following sections the MOSFETs active parts like the pn and MIS junction, are explained in more detail.

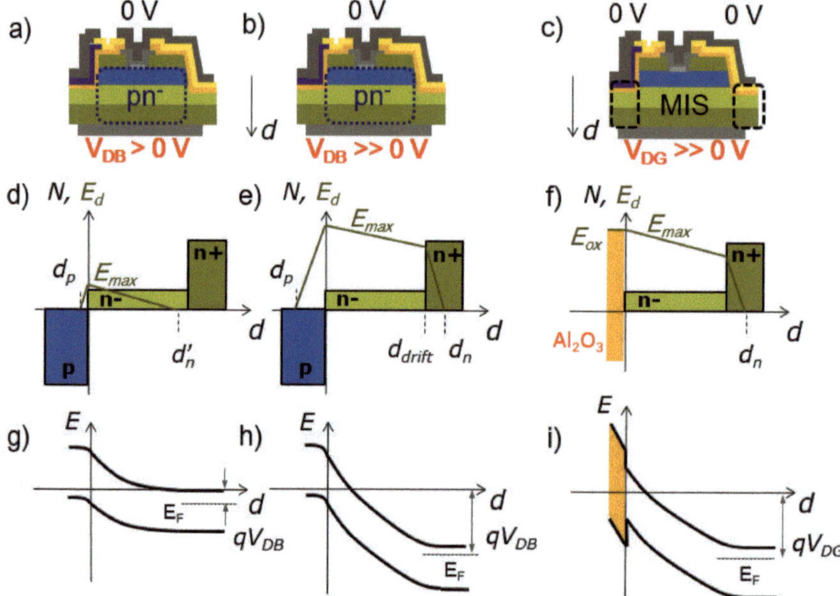

Fig. 2.3: Schematic cross sectional view of a true-vertical MOSFET (a, b, c), with electric field and carrier concentration (d, e, f) and band scheme (g, h, i) of the pn⁻n⁺ (a, b, d, e, g, h) and MIS junction (c, f, i). N refers to the carrier concentration in the n⁻, n⁺ and p-GaN layer, E_d represents the electric field and E the energy. The values d with different indexes are explained in Section 2.3.1

2.3.1 The pn-junction in reverse operation

The pn⁻ body junction is crucial for the device OFF-state, due to its large dimensional contribution to the area of the device. Its lowly doped n⁻ layer is called drift layer and buffers the high voltage during OFF-state operation. Increasing the drift layer thickness (d_{drift}) can increase the possible breakdown voltage. It likewise increases the ON-resistance of the diode and vice versa at similar doping levels. On the other hand the forward conductivity of the junction is important for the avalanche robustness and inverse operation of the MOSFET. In both cases a high current through the junction should not damage the device by the dissipated heat. Thus a low resistance faces a high breakdown voltage. The common way to design a power MOSFET drift region is to use a low/high n-doping profile, which forms a trapezoidal space charge region under high drain voltages as shown in Fig. 2.3 b), e), and h) with the depletion depths d_p and d_n inside the p- and n⁺-GaN, respectively. For correctness the drain-source bias is not directly shown, but the acting voltages V_{DB} and V_{DG} at the respective junction. However, under high voltage conditions the differences between these and V_{DS} are not significant.

The advantage of the n⁻/n⁺ layer stack can be found in the trade-off between ON-resistance and breakdown voltage of the structure. The resistance accompanied with the maximum depletion layer thickness in the n-GaN R_{scr} can be calculated by:

$$R_{scr} = \frac{d_{drift}}{q\mu_{n-}N_{D+n-}} + \frac{d_n}{q\mu_{n+}N_{D+n+}}, \tag{2.5}$$

where the indices indicate the respective layer and its doping. Under low bias conditions the depletion will not reach the highly doped drain layer N_{D+n+} and thus d_{drift} has to be exchanged by the actual depletion depth. The second term associated with the n$^+$ layer disappears since d_n is zero. Generally its resistance contribution is small compared to the drift layer, which is represented by the first term in Equation (2.5).

But before focussing on the discussion of the ON-resistance again, some considerations about the device design for the OFF-state operation are discussed. At high bias conditions $V_{DB} >> 0$ V as shown in Fig. 2.3 b), e) and h) the voltage across the different layers of the structure is distributed as related to

$$V = \frac{qN_{D+n+}}{2\varepsilon_0\varepsilon_{GaN}}d_n{}^2 + \frac{qN_{D+n-}}{2\varepsilon_0\varepsilon_{GaN}}d_{drift}{}^2 + \frac{qN_{D+n+}}{\varepsilon_0\varepsilon_{GaN}}d_nd_{drift} + \frac{qN_{A-}}{2\varepsilon_0\varepsilon_{GaN}}d_p^2 \tag{2.6}$$

in the three differently doped regions of the pn$^-$n$^+$ junction. The maximum electric field E_{max} occurs at the p-GaN/n$^-$-GaN interface of the trapezoidal E-field distribution in the space charge region. It can be calculated by:

$$E_{max} = \frac{qN_{A-}}{\varepsilon_0\varepsilon_{GaN}}d_p = \frac{qN_{D+n+}}{\varepsilon_0\varepsilon_{GaN}}d_n + \frac{qN_{D+n-}}{\varepsilon_0\varepsilon_{GaN}}d_{drift}. \tag{2.7}$$

The build-in field of the diode can be included in this equation but is neglected here, due to its insignificance when high voltages are applied in OFF-state. The maximum electric field at the p-GaN/n-GaN interface can be calculated using Equation 2.7. However, in this case the values of d_p and d_n have to be known. The most convenient way to obtain them is to reformulate (2.7) for d_p and insert the term in Equation (2.6) to obtain d_n as a function of the voltage. The quadratic relation can then easily be solved for d_n and E_{max} as well as d_p, which is not shown here. The same structure under low bias conditions $V_{DB} > 0$ V is shown in Fig. 2.3 a). In this case the space charge region does not reach the n$^+$ layer. A triangular shaped E-field distribution with a depth d_n' is formed inside the drift layer, see Fig. 2.3 a), d) and g). The Equation (2.6) and Equation (2.7) have thus to be modified. By setting d_n to zero and changing d_{drift} to d_n' the formulas are transferred into the relations for a classical pn-junction for the voltage:

$$V = \frac{qN_{D+n-}}{2\varepsilon_0\varepsilon_{GaN}}d_n'{}^2 + \frac{qN_{A-}}{2\varepsilon_0\varepsilon_{GaN}}d_p^2 \tag{2.8}$$

and the maximum electric field:

$$E_{max} = \frac{qN_{A-}}{\varepsilon_0\varepsilon_{GaN}}d_p = \frac{qN_{D+n-}}{\varepsilon_0\varepsilon_{GaN}}d_n'. \tag{2.9}$$

Fig. 2.4 shows the dependence of the depletion depth and maximum operation voltage of the device on the drift layer doping concentration N_{D+n-} for two different values of d_{drift}. A breakdown field strength of 2 MV/cm is used in the calculation. Whereas the literature value is about 3.3 MV/cm for avalanche breakdown (table 1.1). In Fig. 2.4 a) the transition between the triangular space charge region is linked to the saturation of the depletion depth at the drift layer thickness due to the high doping concentration of $5 \cdot 10^{18}$ cm^{-3} in the n$^+$ layer. For very low drift layer doping concentrations a trapezoid shaped space charge region occurs already at very low bias. In part b) of the figure the transition can be related to the inflexion point of the curve. From both figures it becomes obvious, that a doping concentration above $5 \cdot 10^{16}$ cm^{-3} is of disadvantage for a high breakdown voltage. Even for low voltage devices with thin drift layers high drift layer doping reduces strongly the achievable breakdown voltage of the MOSFET.

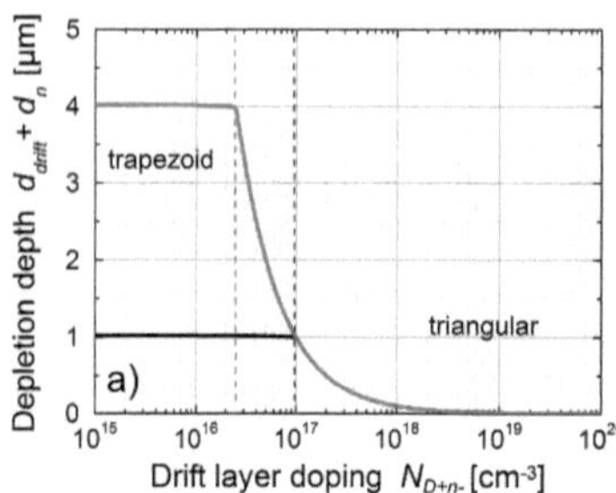
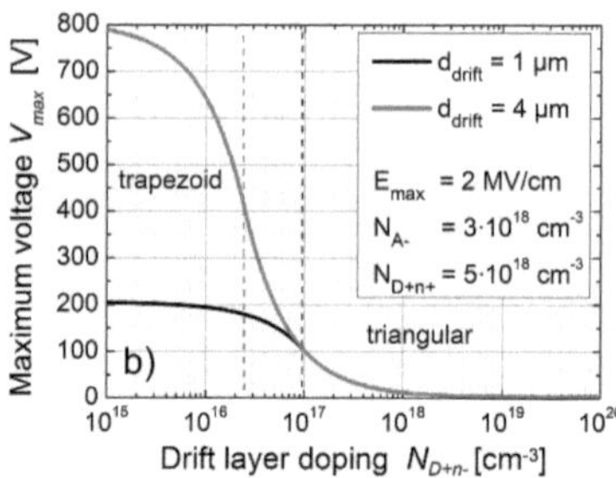

Fig. 2.4: Depletion depth (a) and maximum breakdown voltage versus drift layer doping concentration N_{D+n-} for different drift layer thicknesses (b). A maximum electric field of 2 MV/cm is assumed. The regions of triangular and trapezoid space charge region are separated by broken lines with the respective colour.

In the following the influence of the p-GaN or dielectric layer is discussed. On the one side, in case of high N_{A-} (abrupt junction) d_p is very small and the voltage drop across the p-GaN depletion layer is negligible. On the other side, for low N_{A-} the voltage drop within the p-GaN layer can be significant and results in a higher overall voltage drop over the structure. The associated trade-off is the extension of the space charge region into the p-GaN layer. If the depletion within the p-GaN over comes the layer thickness a punch-through occurs and a high current from the source side can potentially destroy the device. It is indispensable to avoid this situation.

As can be seen from Equation (2.5) and Equation (2.6) the breakdown voltage scales in a quadratic way, whereas the resistivity R_{scr} scales linear with the drift layer thickness. For the optimization of the drift layer design the question arises, which design gives a suitable trade-off between the resistance and a certain target breakdown voltage. This interplay is shown in Fig. 2.5 with the two different cases of triangular and trapezoid space charge region in b). The reduction ratios of the achievable breakdown voltages and resistances referring to the maximum values in the triangular space charge region case (without reduction) are shown in a). When the drift layer thickness is reduced the equivalent resistance of the space charge region drops linear and the breakdown voltage drops in a quadratic way with the ratio of the thickness reduction.

The trade-off drift layer thickness can be specified dependent on the desired device parameters. One possible trade-off between the decrease of the resistance and the reduction of the breakdown voltage is achieved at a drift layer thickness of 70 % compared to full drift layer depletion. As before a breakdown field strength of 2 MV/cm was chosen. The breakdown voltage is reduced by 8 % in this case. However, the corresponding resistance drops by 30 %.

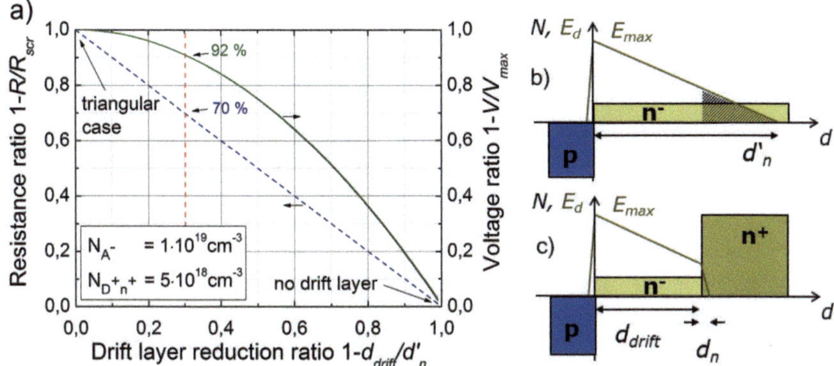

Fig. 2.5: a) Resistance and breakdown voltage reduction ratio versus drift layer reduction ratio, relatively to the triangular case shown in b). The data for two different drift layer doping concentrations ($1 \cdot 10^{16}$ cm^{-3} and $3 \cdot 10^{16}$ cm^{-3}) is overlying each other. The resistance and breakdown voltage drops to very small values at 1-$d_{drift}/d'_n = 1$, which means, the drift layer thickness is zero. At 1-$d_{drift}/d'_n = 0$ it represents the transition from trapezoid (c) to triangular case (b) with a drift layer thickness larger than the depletion width. A possible trade-off is marked by the red broken line. The schematics b) and c) show the electric field and voltage re-distribution with limited drift layer thickness, similarly to Fig. 2.3 b), e) and h). The shaded area in b) represents the voltage gain at constant E_{max} for the triangular case.

This assumption holds true independently of the drift layer doping (N_{D+n-}) as long as both junctions are abrupt ($N_{D+n-} \ll N_{A-}$ and $N_{D+n-} \ll N_{D+n+}$).

As discussed previously a maximum electric field of 3.3 MV/cm is stated for GaN. This value typically relates to the inset of avalanche multiplication. However, various other leakage current mechanisms have been investigated in literature, such as surface leakage, variable range hopping, Poole-Frenkel (trap assisted) leakage and space charge limited current. A more detailed discussion on each mechanism can be found elsewhere [61, 62, 63]. In table 2.1 the mathematical expressions and dependencies are summarized for these potential leakage mechanisms.

Tab. 2.1: Names and dependencies for different leakage current mechanisms related to bulk and surface contributions. u denotes the circumference of the measured test structure.

Name	Dependencies
Surface leakage	$I \sim E, I \sim u$
Variable range hopping	$ln(I) \sim E, I \sim A, I \sim \text{TDD}, ln(I) \sim T^{\frac{1}{4}}$
Poole-Frenkel leakage	$ln(I) \sim \sqrt{E}, I \sim A, I \sim N_T, ln(I) \sim T$
Space charge limited	$I \sim E^m (m > 1), I \sim A, I \sim d_{drift}^{-3}$

From the different dependencies of the leakage current on temperature, geometry and the electric field the contributions can be discriminated. It should be noted that for a triangular shaped space charge region, the peak electric field at the pn-junction scales with the square root of the voltage. In the trapezoidal case it scales linearly with the voltage. A surface leakage path along the isolation sidewall is likely when the leakage current is not scaling with the area but with the circumference of the DUT. In particular, when a rather linear behaviour can be observed. Variable range hopping is often linked to threading dislocations in literature. For the here used geometries the amount of dislocations is still statistically high on a single device. The current thus scales with the TDD, which are typically present with different concentrations on various substrate materials. Trap assisted leakage such as the Poole-Frenkel mechanism are depending on the defect density and their energetic distribution within the band gap. Another reported mechanism is space charge limited leakage, which shows a dependence with a potential coefficient larger than unity. For strongly localized leakage paths e.g. along dislocations or on defects on the device sidewall the high current density results in a local temperature increase. In these regions the current density can rise additionally due to the positive temperature coefficient of the leakage mechanism. This results in a thermal runaway, under high current densities and insufficient thermal conduction of the heat to the surroundings. The corresponding observation is a rather soft breakdown of the device, which can be destructive if the dissipated power is too large.

2.3.2 The gate trench MIS-structure in OFF-state

Similar estimations as for the pn^-n^+ junction can be made for the MIS-structure in OFF-state. It behaves certainly like an abrupt junction. Also here, the field drop within the dielectric can be significant for thick dielectrics d_{ox} e.g. used under a field plate. Additionally low-k dielectrics have to withstand a higher electric field according to the factor $\varepsilon_{GaN}/\varepsilon_{ox}$. The reduced ε_{ox} will further increase the voltage drop in the layer. The latter effect can be used beneficially, especially in pseudo-vertical devices, due to the relatively thin drift layer. In a similar way as for the equations for the pn^- structure the gate MIS-structure is described with

$$V = \frac{qN_{D^+ n^+}}{2\varepsilon_0\varepsilon_{GaN}}d_n{}^2 + \frac{qN_{D^+ n^-}}{2\varepsilon_0\varepsilon_{GaN}}d_{drift}{}^2 + \frac{qN_{D^+ n^+}}{\varepsilon_0\varepsilon_{GaN}}d_n d_{drift} + \frac{\varepsilon_{GaN}}{\varepsilon_{ox}}E_{max}d_{ox} \qquad (2.10)$$

for the voltage and

$$E_{max} = \frac{\varepsilon_{ox}}{\varepsilon_{GaN}}E_{ox} = \frac{qN_{D^+ n^+}}{\varepsilon_0\varepsilon_{GaN}}d_n + \frac{qN_{D^+ n^-}}{\varepsilon_0\varepsilon_{GaN}}d_{drift} \qquad (2.11)$$

for the peak electric field. The electric fields at the transition from GaN to dielectric can be related to each other by using Gauss's law with $E_{ox}\varepsilon_{ox} = E_{max}\varepsilon_{GaN}$ as described in (2.11). From Equation (2.7) and (2.11) it can be seen, that the only difference between the pn junction and MIS-structure relates to the voltage portion associated to the p-GaN and dielectric, respectively. In Fig. 2.6 the voltage to reach E_{max} for different structures is shown. Notice that a reduced N_{A^-} acts similarly as if a thick and/or low-k dielectric is used. In both cases an offset is added to the V_{max} scale in Fig. 2.6, which represents the additional voltage drop on the p-GaN or dielectric side of the junction.

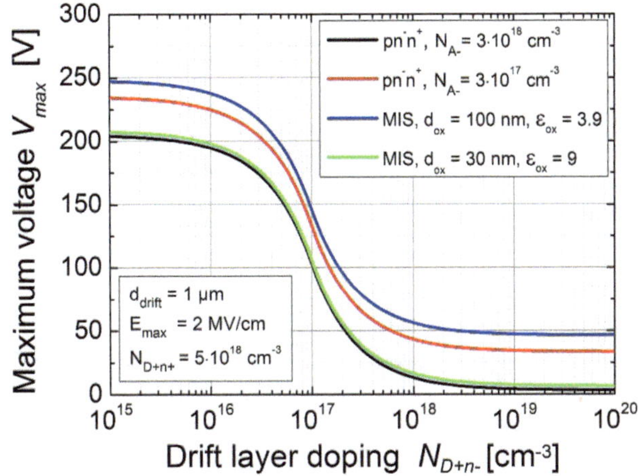

Fig. 2.6: Maximum voltage, which can be applied to the structure under assumption of a breakdown field of 2 MV/cm and $d_{drift} = 1$ µm versus drift layer doping concentration. The influence of different p-GaN dopings and dielectrics is shown, whereas in the case of $d_{ox} = 30$ nm Al_2O_3 with a $\varepsilon_{ox} = 9$ and for $d_{ox} = 100$ nm SiO_2 with $\varepsilon_{ox} = 3.9$ was assumed.

2.3.3 Dimensional constraints and field plates

In [64] simulations of the electric field in the area of the gate trench foot were conducted, showing a high field strength on the bottom corner of the trench at the transition from vertical edge to trench foot. Additionally, it is shown that the electric field strength on the bottom of the trench increases with a smaller gate foot breadth. The higher field strength leads to stronger leakage through the dielectric and to earlier electrical breakdown of the structure. To avoid these effects certain design constraints are meaningful.

As described previously, the area consumed by the device design is usually kept as small as possible in order to reduce the device footprint and thus the cost per device on the wafer as well as its $R_{DSon} \cdot A$ figure of merit. Additionally, a smaller junction area reduces leakage contributions. Connection lines on the surface get short and benefit from low parasitics. Challenges are likewise the contact formation to the functional layers with small but low resistive contacts, highly conductive metal lines and thermal coupling to cooling structures. The isolation of devices on the wafer and die can have various possible implementations. Due to difficult implantation in GaN and hardly realizable negative, undercutted isolation edges usually field plates on positively sloped isolation edges are applied [65]. In principle, the field plate operates similar like the device gate in OFF-state, but usually the voltage on the field plate is kept constant, below the threshold voltage. The metal plate is often connected to the source by a higher metallization layer. Due to the formation of a depletion layer under the plate it improves the buffering of field peaks at the pn⁻ junction, corners and interfaces in vicinity to the surface. To reach even higher breakdown voltages than for the gate, a second dielectric layer and/or low-k dielectric can be used. It also acts as second passivation. Simulation studies were made on similar devices in [66]. Oka et al. showed a successful implementation of a field plate for true vertical MOSFETs, see [67].

2.4 Static ON-state and switching considerations

In this section the ON-state and the static behaviour of the device are discussed. First in Section 2.4.2 the contributions of different device parts to the entire R_{DSon} are explained and a series resistance model of the device is presented. Next, possibilities for the estimation of intrinsic channel parameters such as the mobility shall be discussed. Linked to that is also the derivation of parameters from the threshold and subthreshold properties as well as the body bias effect on the threshold voltage.

2.4.1 The pn-junction in forward operation

In the following the forward operation of the pn⁻ diode is focused. The discussion targets the diode as part of the MOSFET, as well as single extrinsic diodes, which can separately be used as power devices. In order to model the behaviour of the diode an additional series R_s and parallel resistance R_p is included. The current into the diode I_{pn} can then be described by

$$I_{pn} = I_{Diff} + I_{b-b} + I_{SRH} + \frac{V_{pn}}{R_p}, \qquad (2.12)$$

where I_{Diff}, I_{b-b} and I_{SRH} describe the current contributions of the diffusion current, band to band recombination current and the trap assisted Schockley-Read-Hall recombination current in the space charge region, respectively. A photo current generated under illumination is neglected. A parallel current I_p is described by the term V_{pn}/R_p. V_{pn} describes the voltage on the junction. Further an external voltage can be defined, which includes the voltage drop on a series resistance. The respective equivalent circuit is shown in Fig. 2.7 a). The series resistance R_s can include bulk resistances of the quasi neutral regions or the substrate. But of particular importance are contact or lead resistances. The external voltage is then described with

$$V = V_{pn} + I_{pn}R_s. \qquad (2.13)$$

A representatin for the equivalent circuit is given in Fig. 2.7. For the different current contributions described by Equation (2.12) a classical formulation is used, which can be found e.g. in [42], [68] and [69].

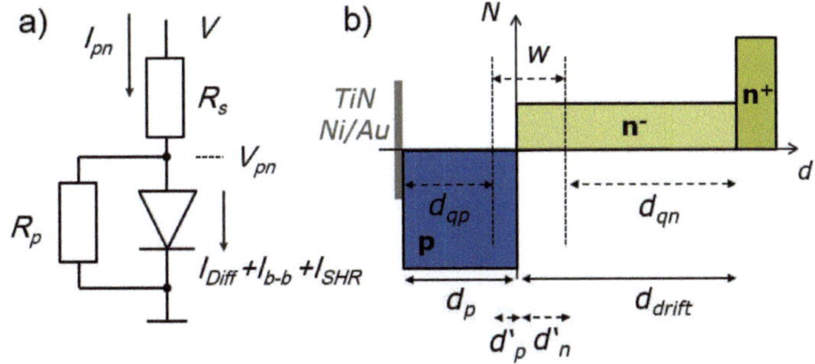

Fig. 2.7: Equivalent circuit (a) after Equation (2.12) and Equation (2.13) with the intrinsic diode, parallel and series resistance; In b) the carrier concentrations are shown over the depth d of the diode. The respective values d, d', d_q and w are shown according to the equations in this section.

The diffusion current can be described by:

$$I_{Diff} = q \left(\frac{D_n \delta n_p}{L_n} coth \left(\frac{d_{qp}}{Ln} \right) + \frac{D_p \delta p_n}{L_p} coth \left(\frac{d_{qn}}{Lp} \right) \right) \left(exp \left(\frac{qV_{pn}}{kT} \right) - 1 \right), \quad (2.14)$$

where the first bracket denotes the saturation diffusion current I_{SDiff}. The values D_n, D_p, L_n and L_p refer to the diffusion coefficients and diffusion lengths of electrons and holes, respectively. D_n and D_p can be calculated with the Einstein relation:

$$D_{n,p} = \mu_{n,p} \frac{kT}{q}, \quad (2.15)$$

whereas L, D and the carrier lifetime τ are connected by:

$$L = \sqrt{D\tau}, \quad (2.16)$$

The length of the quasi neutral regions for the n and p side are described in a first approximation for thick drift layers by $d_{qn} = d_{drift} - d'_n$ and $d_{qp} = d_p - d'_p$. In forward operation of the diode the thickness of the space charge region w is much smaller than in OFF-state operation. It can not reach the n$^+$ drain layer. This case is represented by Fig. 2.7. Equation (2.14) describes the diffusion current in a general way including low and high injection by the excess carrier concentrations δn_p and δp_n at the edge of the space charge region in p and n quasi neutral region, respectively. A closer discussion and deduction is presented in the appendix A.1.

The band to band recombination current is described by:

$$I_{b-b} = q n_i^2 bw \left(exp \left(\frac{qV_{pn}}{kT} \right) - 1 \right), \quad (2.17)$$

where b represents the radiative recombination coefficient of GaN which is reported to be $1.1 \cdot 10^{-11}$ cm^{3}s^{-1} in [70]. It was determined from optical investigation on GaN on sapphire. Another value is reported in [71] with $4.7 \cdot 10^{-11}$ cm^{3}s^{-1}, which is based on calculations for the ideal material. Typical values for Si and SiC are about $1 \cdot 10^{-14}$ cm^{3}s^{-1} [72] and $1 \cdot 10^{-12}$

cm^3s^{-1} [73], respectively. Thus, the band to band recombination probability is at least one order of magnitude higher in GaN. w represents the space charge region thickness by $w = d'_n + d'_p$. The Schockley-Read-Hall (SRH) recombination current can be described by:

$$I_{SRH} = \frac{qw_{eff}c_nc_p(pn - n_i^2)}{c_n\left(n + n_i exp\left(\frac{E_T - E_i}{kT}\right)\right) + c_p\left(p + n_i exp\left(\frac{E_i - E_T}{kT}\right)\right)},\qquad(2.18)$$

with the effective recombination width w_{eff}, and the capture rates for electrons and holes (c_n and c_p). They are inverse to the respective carrier lifetimes (τ_n and τ_p) and can be calculated by:

$$c_{n,p} = v_{thn,p}\sigma_{n,p}N_T = 1/\tau_{n,p}.\qquad(2.19)$$

The SRH current describes the recombination current over trap states inside the band gap. A certain trap concentration N_T at an energy E_T is assumed. Often Equation (2.18) is reformulated in order to obtain a maximum estimation for the Schockley-Read-Hall recombination current. This is typically given with

$$I_{SRH} = q\frac{n_i}{2\tau_{rc}}w\left(exp\left(\frac{qV_{pn}}{2kT}\right) - 1\right),\qquad(2.20)$$

where τ_{rc} represents the recombination lifetime for carriers with an ambipolar-equal midgap trap, an averaged constant recombination in the space charge region and the assumption of $w_{eff} = w$ [69]. In this case the carrier concentrations n and p can be described by:

$$n = p = n_i\left(exp\left(\frac{qV_{pn}}{2kT}\right) - 1\right).\qquad(2.21)$$

For the calculation of d'_n and d'_p please see appendix A.1.

A rough estimation of the diodes "quality" is usually made by the Schockley equation, which is only valid for the exponential part of the $I(V)$ curve of the diode and is typically used for small signal fits of the characteristics. It can be described with:

$$I_{pn} = I_{SDiff}\left(exp\left(\frac{qV_{pn}}{n_{if}kT}\right) - 1\right),\qquad(2.22)$$

so that n_{if} relates to the slope of the exponential part in logarithmic scale, which is limited to a maximum of approximately one decade current per 60 mV voltage. In this case n_{if} equals unity. It is called ideality factor of the diode. From Equation (2.12), Equation (2.14), Equation (2.17) and Equation (2.20) it can be seen, that its value is valid in the range of 1 to 2. However, if a resistive component (e.g. R_s) is added to the model this approximation shows strong discrepancies. Although physically not very valuable ideality factors larger than 2 are obtained in then.

2.4.2 Resistance contributions

To evaluate the quality of the device gate trench and the interface between the GaN layers and the dielectrics, the intrinsic channel properties such as mobility and carrier density can be used. However, parasitic series resistances and trapping effects can deteriorate the values obtained by the direct calculation from measurements. In order to extract more valuable parameters a model is used, which includes series resistance contributions as shown in Fig. 2.8.

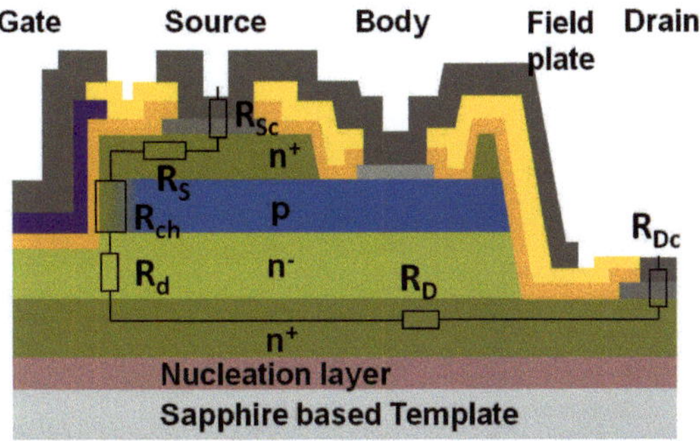

Fig. 2.8: Schematic cross section of a pseudo-vertical MOSFET with its series resistance contributions to the overall ON-resistance

The channel resistance R_{ch} can be calculated by

$$R_{ch} = R_{DSon} - (R_{Dc} + R_{Sc} + R_S + R_D + R_d) \tag{2.23}$$

and by extracting the MOSFETs R_{DSon} and measuring the other values on additional test structures. Values, which are hardly accessible by measurements can be estimated by calculation from the design parameters and dimensions. Typical cooperative structures are linear or round transfer length method (TLM) structures or MIS-structures, which are closer described in Section 3.4.3. The n$^+$ drain layer resistance R_D, the source n$^+$ layer resistance R_S and the contact resistances (R_{Dc} and R_{Sc}) can be obtained from TLM-structures. R_d can be calculated by dimensional constrains and with the specific material resistance. Evaluating capacitance versus voltage ($C(V)$) measurements in combination with the sheet resistances obtained from TLM-structures grants access to the doping concentrations and mobilities in the respective layers. The sheet resistance in wide band gap material is dominated by majority carriers, thus the mobility can be estimated by:

$$\mu = \frac{1}{qnR_{sh}d}. \tag{2.24}$$

The metal line resistances are neglected in the shown modelling. In Fig. 2.9 the effects of increased source-sided and drain-sided parasitic resistances on the transfer and output characteristics are shown. An increased source-sided resistance deteriorates the device ON-behaviour much more than drain sided contributions. It increases not only the R_{DSon}, but leads also to a reduction of g_m and I_{Dsat}. It appears as a dynamic shift of the device transfer characteristics during increasing drain current (see Fig. 2.9 a) and b)).

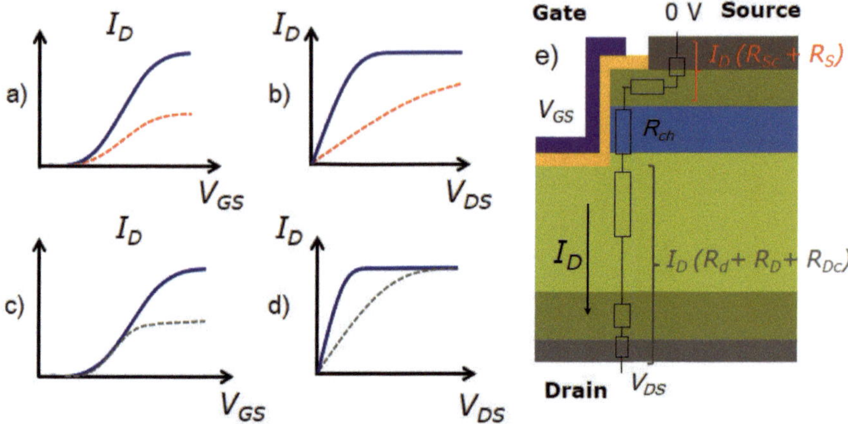

Fig. 2.9: The influence of parasitic voltage drops on source sided (a, b) and drain sided (c,d) resistances on the transfer ($I_D(V_{GS})$) and output characteristics ($I_D(V_{DS})$) are shown. The full lines represent the schematic transfer and output characterisitcs, whereas the broken lines denote the degraded characteristics. In e) the voltages contribution from the resistive contributions $I \cdot R$ are denoted with their respective colour code.

An increased drain-sided resistance contributes to the R_{DSon} only, i.e. a higher drain-source voltage has to be applied to achieve the same drain current (Fig. 2.9 c) and d)). The relations between the external voltages applied to the gate and drain terminal and the intrinsic channel voltages are given by

$$V_{GS,ch} = V_{GS} - I_D(R_S + R_{Sc}) \tag{2.25}$$

and

$$V_{DS,ch} = V_{DS} - I_D(R_d + R_D + R_{Dc} + R_S + R_{Sc}). \tag{2.26}$$

The different contributions according to the equations are shown in Fig. 2.9 e). Once R_{ch} is deduced, the complete device can be modelled. $V_{DS,ch}$ and $V_{GS,ch}$ represent in this case the voltages at the intrinsic channel and V_{DS} and V_{GS} the voltages applied to the outer device terminals. With that common model of the MOSFET the channel mobility can be estimated [37]. This topic is revisited in the next section.

2.4.3 Device model and channel mobility

The channel of the device can be modelled using (2.27) representing the "EKV model" developed by Enz, Krummenacher and Vittoz [74], where b is the body capacitance factor $b = 1 + C_D/C_{ox}$ with C_D as the depletion layer capacitance. It represents the ratio of voltage controlling the channel and depletion layer charge density. The expression for the drain current is:

$$I_D = 2b\mu_{ch}C_{ox}\frac{W_G}{L_G}\left(\frac{kT}{q}\right)^2\left(ln^2\left(1+exp(G)\right)-ln^2\left(1+exp\left(G-\frac{qV_{DS}}{2kT}\right)\right)\right), \tag{2.27}$$

the value G is here represented by the term:

$$G = \frac{q(V_{GS} - V_{th})}{2bkT}.$$

(2.28)

The influence of V_{BS} is not included here, please see [68]. The EKV model combines all ON-operation modes including the subthreshold operation. Different device operation regimes are discussed in [68]. The subthreshold regime is developed by the first element of the Taylor series at subthreshold conditions shown in Equation (2.29). The linear and quadratic approximation are shown with (2.30) and (2.31), respectively. A more detailed deduction can be found in appendix A.2.

- Subthreshold region ($V_{GS} < V_{th}$, $V_{DS} > 0$ V)

$$I_D = 2b\mu_{ch}C_{ox}\frac{W_G}{L_G}\left(\frac{kT}{q}\right)^2\left(1 - exp\left(\frac{-qV_{DS}}{kT}\right)\right)exp\frac{q(V_{GS} - V_{th})}{bkT}$$

(2.29)

- Linear region ($V_{GS} > V_{th}$, $V_{DS} < V_{GS} - V_{th}$)

$$I_D = \mu_{ch}C_{ox}\frac{W_G}{L_G}\left(V_{GS} - V_{th} - \frac{bV_{DS}}{2}\right)V_{DS}$$

(2.30)

- Quadratic region ($V_{GS} > V_{th}$, $V_{DS} > V_{GS} - V_{th}$)

$$I_D = \mu_{ch}C_{ox}\frac{W_G}{L_G}\left(\frac{(V_{GS} - V_{th})^2}{2b}\right)$$

(2.31)

The different operation regions are shown in Fig. 2.10. The schematic transfer characteristics (a) are shown for two different V_{DS} in linear and logarithmic I_D scale, whereas the schematic output characteristics (b) are shown with three different V_{GS}. As shown, the saturation current is increasing with a square dependency when V_{GS} and thus $V_{GS} - V_{th}$ is linearly increased.

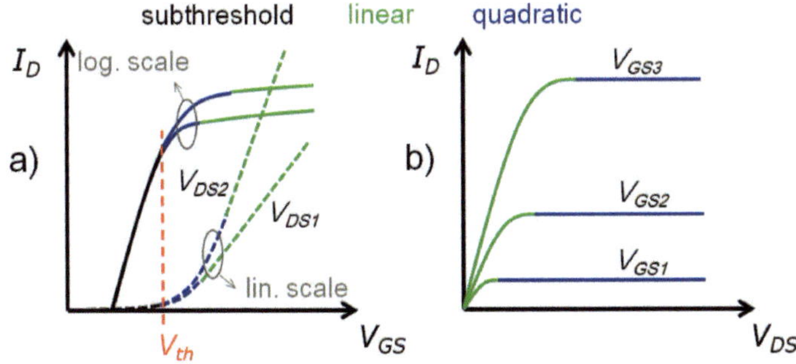

Fig. 2.10: The schematic MOSFET transfer characteristics (a) for two different drain-source voltages V_{DS1} and V_{DS2} in linear and logarithmic scale and output characteristics (b) for three different gate source voltages V_{GS1}, V_{GS2} and V_{GS3} of a MOSFET are shown. The different operation regimes, i.e. linear, quadratic and subthreshold region are colour coded.

Prior to device fabrication, the described model enables the estimation of device characteristics including a simple series resistance correction by Equation (2.25) and Equation (2.26) in case the specific device properties like channel mobility μ_{ch}, oxide capacitance C_{ox}, threshold voltage V_{th} and body capacitance factor b can be estimated. However, in many cases these values cannot be pre-assumed easily and the application of the model is limited to a fit to the characteristics after the device processing and to subsequent deduction of the interesting parameters.

With

$$\mu_{FE} = \frac{L_G g_m}{W_G C_{ox} V_{DS}} \tag{2.32}$$

and

$$\mu_{eff} = \frac{L_G}{W_G C_{ox} R_{DS}(V_{GS} - Vth)} \tag{2.33}$$

the channel mobility can be extracted from the transfer (by g_m) and output characteristics (by R_{DS}), directly. However, they do not include external device series resistances such as contact, layer or metal line resistances. Moreover, the usage of μ_{FE} does not include a correction for the field effect on the mobility, it is thus typically referred to as field effect mobility. Generally, both obtained values underestimate the actual channel mobility without correction for series resistances. Whereas the results for the effective mobility μ_{eff} give higher values [75]. Other uncertainties can arise from geometrical effects resulting from the device layout, which are not included in the model. For the here shown structures the last are neglected, due to high channel width to edge ratios.

2.4.4 Threshold voltage and subthreshold slope

The threshold voltage (V_{th}) of a MOSFET can be calculated using the expression:

$$V_{th} = \frac{\Phi_{MS}}{q} - \frac{Q_{eff}}{C_{ox}} + 2\phi_F + \gamma\sqrt{2\phi_F - V_{BS}}. \tag{2.34}$$

The substrate control factor is represented by γ and defined by

$$\gamma = \frac{\sqrt{2q\varepsilon_0\varepsilon_{GaN}N_{A^-}}}{C_{ox}}, \tag{2.35}$$

where ϕ_F is the Fermi level with respect to the intrinsic Fermi energy. Φ_{MS} is the metal-semiconductor work function difference, C_{ox} is the oxide capacitance and Q_{eff} reflects all charges effectively located at the dielectric/GaN interface under threshold conditions. Classically only fixed charges (Q_{fix}) are considered, as further discussed in Section 2.4.5. The body-source voltage is represented by V_{BS}. The flat band voltage V_{FB} is substituted by a similar term as shown in (2.36) and can thereby be calculated according to

$$V_{FB} = \frac{\Phi_{MS}}{q} - \frac{Q_{FB}}{C_{ox}}. \tag{2.36}$$

Q_{FB} accounts for all charges effectively located at the dielectric/GaN interface at flat band conditions, thus probably only a portion of Q_{eff}. In most cases also here only fixed oxide charges are considered, due to the low bias applied at flat band conditions and the fast response of interface traps. Under these conditions Q_{fix} is constant and equal at V_{FB} and V_{th}. However, in wide band gap material this classic understanding is not always sufficient. With Equation (2.34) the threshold voltage can be estimated before the actual device is built, often N_{Mg} is used instead of N_{A-} to estimate V_{th}. But, in Section 2.1 several reasons for the deviation of the ionized acceptor concentration compared to the incorporated Mg concentration are discussed, which can lead to unexpected threshold voltage estimations based on the incorporated Mg content.

If $V_{GS} < V_{th}$ with a drain current higher than the leakage level the device is operated in the so called subthreshold regime. The current-voltage behaviour in this region is very similar to the low bias forward operation of a pn-diode or a bipolar transistor. The gate-source voltage needed to increase the drain current by a factor of 10 is usually called subthreshold swing. Its value can be estimated using

$$S_{sth} = \frac{dV_{GS}}{dlog(I_D)} = \frac{ln(10)kT}{q}\left(1 + \frac{C_D + C_{it}}{C_{ox}}\right), \qquad (2.37)$$

which results from Equation (2.29), where C_D is the depletion capacitance within the semiconductor and C_{it} the interface trap capacitance, which is effectively parallel to C_D. The body capacitance factor $b = 1 + (C_D + C_{it})/C_{ox}$ is extended by an interface trap capacitance. It represents the ratio between the voltage used to populate and deplete the channel and the voltage used to charge the depletion region and interface states. The minimum swing at room temperature is ≈ 60 mV per decade drain current (mV/dec). It can be reached when a high C_{ox} compared to $C_D + C_{it}$ is provided and when $C_{it} = 0$. It means that the entire voltage is used to control the channel. Neglecting interface traps, a higher body layer doping of the p-GaN leads to a higher C_D and hence to a higher subthreshold swing. Contrarily, lower p-doping decreases the gate coupling and decreases C_D, but bears the risk of punch-through under high voltage operation. Additionally good ohmic contacts to low doped p-GaN body layers are hardly achievable. Thus a reasonable compromise has to be made considering additionally the absolute threshold voltage.

Typically a V_{th} of 2 - 5 V is targeted. It should ensure safe turn-OFF at $V_{GS} = 0$ V and low gate overdrive to switch the device to a fully ON-state. A doping concentration in the range of $1 \cdot 10^{18}$ cm^{-3} - $1 \cdot 10^{19}$ cm^{-3} is usually used.

2.4.5 Interface and dielectric trap states in wide band semiconductors

The presence of traps within a semiconductor device, in particular in the gate stack can have severe influences on the device properties and its performance. Typically in small band gap material the different trap types are ordered by their behaviour under different bias conditions:

- spatially stationary or mobile
- fixed in polarity and charge or re-chargeable

In wide band gap material this classification remains. In the same way the classification for their position inside a typical layer stack can be resumed e.g. for a MIS-structure or a

MOSFET gate stack:

- bulk dielectric traps (Q_{ox})
- dielectric/semiconductor interface trap states (Q_{it})
- traps close to the interface but inside the dielectric (border traps Q_{bt})
- bulk traps in the semiconductor material (Q_b)

However, the resulting properties are different in GaN based devices compared to silicon structures, due to the wide band gap of this semiconductor. A short review on these points is thus necessary.

Mobile traps like potassium or sodium ions are not present in significant concentrations in a modern clean room based laboratory and technology. Additionally, diffusion in III-N material is gernerally impeded [76]. Thus, they are not further discussed here and considered as negligible. Border traps, which are within 3 nm to the GaN/dielectric interface inside the dielectric can interact with interface states and the semiconductor by tunneling. Their signature is similar to deep traps located directly at the interface. Unless deeper investigations on the trap type are performed a descrimination from deep interface traps is thus hardly possible. In the following they are not separately discussed. However, fixed and rechargeable traps states are typically present in significant concentrations and shall be focused here. There energetic and spatial distribution is schematically shown in Fig. 2.11. The dynamics of traps spread over a much larger time scale, since trap states can be located energetically much deeper inside the band gap. For electrons the time constant for emission from trap states to the conduction band τ_{tn} can be approximated by [77]:

$$\tau_{tn} = \frac{1}{N_C v_{thn} \sigma_n} exp\left(\frac{E_C - E_T}{kT}\right). \tag{2.38}$$

At room temperature the emission time constant of a shallow trap state (e.g. $E_C - E_T = 100$ meV) is in the nanosecond range. Whereas, for a deep trap with $E_C - E_T > 1$ eV the emission time constant is longer than one day. Even at elevated temperature midgap states relaxation times are very long. It is thus hard to find a definition for fixed charge traps. Typical measurement intervals expand over some ten seconds, minutes or even hours. States, which do not react in this time frame are considered usually as fixed, although they are versatile on long time scale. Trap states inside the oxide layer are usually considered as fixed in a similar way. They lie deep inside the dielectrics band gap and thus have very long relaxation times. A variation of the oxide thickness can help to discriminate between bulk oxide and interface trapped charges [78]. The latter are typically present with decaying tails from the band edges in GaN [79, 80].

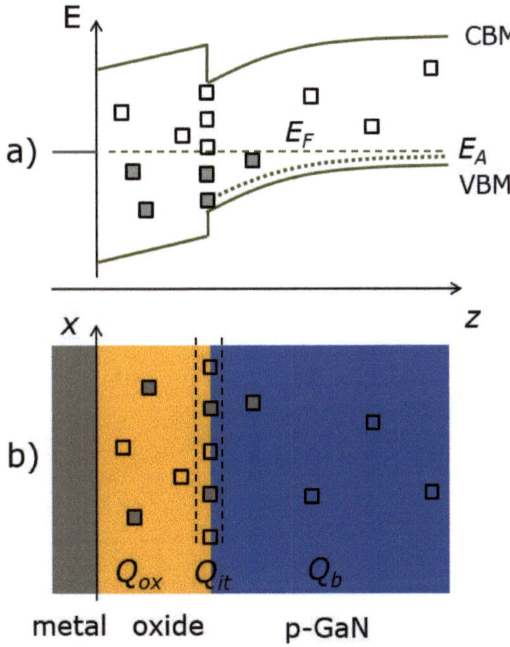

Fig. 2.11: In a) a schematic band structure E(z) and in b) a possible spacial distribution of trap states (b) is shown. Trap states at the interface , in the p-GaN bulk layer and inside the dielectric are denoted by Q_{it}, Q_b and Q_{ox}, respectively. Filled squares denote electron-filled trap states and transparent squares represent empty trap states.

Although they are energetically more localized, the evaluation of bulk semiconductor charges is more difficult, due to their spatial and energetic distribution. Further superposition with effects of Q_{it} impedes their analysis. Possible methods to evaluate the location and energetic signature of these states are deep level transient spectroscopy (DLTS) [81] or photo-induced current transient-spectroscopy (PICTS) [82] on special test structures. In Si with high purity and crystal quality deep level defect concentrations are insignificant. Mostly and solely significant are interface traps. The trap analyses on e.g. MIS-structures is thus not very problematic. During the characterisation of new wide band gap materials with less purity and crystal quality a usual way is to project all charge to the interface plane. Finally, only a trap density, which is effectively located at the dielectric/semiconductor interface is given. A similar strategy is used in this work. The obtained density still reflects the total contribution of traps at the actual operation conditions, but gives limited information for the real amount and spatial distribution of the traps. Further information can be obtained by extended characterisation, or if one kind of traps can be neglected. However, this is mostly not the case.

2.4.6 The body bias effect

In this section the extraction of the actual active acceptor concentration in a space charge region is presented. In Fig. 2.12 the schematic conduction band landscape under the gate of the MOSFET is shown. $E_{F(n+)}$ and $E_{F(p)}$ represent the bulk Fermi levels of the p- and n $^+$-GaN layer, respectively.

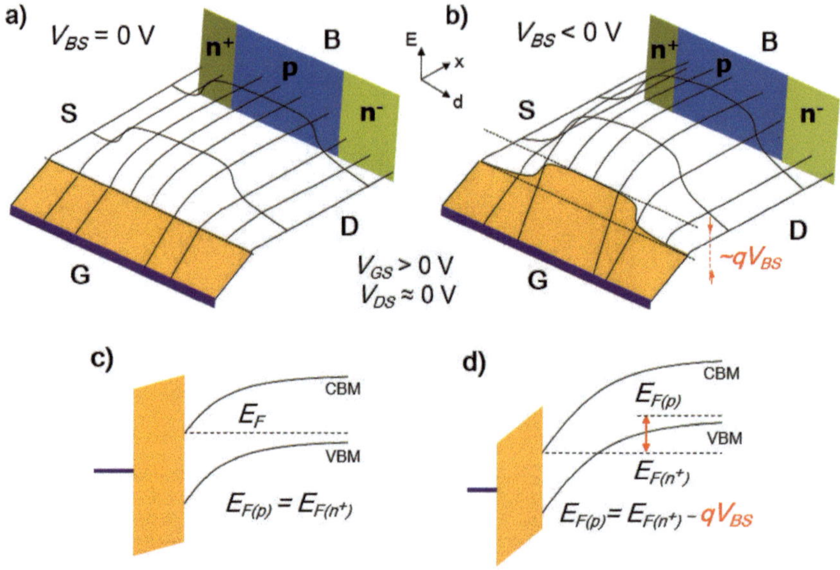

Fig. 2.12: Schematic conduction band landscape in the vicinity of the gate of the MOSFET in depletion/weak inversion at low V_{DS} without (a) and with body bias (b). The band structure under the gate in the p-GaN is shown in c) and d) under similar conditions. The structure is shown with similar V_{GS}. A cross sectional view of the band structure from the gate into the p-GaN is shown in c) and d). The additional potential barrier for electrons injected from the source side into the channel is formed by qV_{BS}.

Under normal operation conditions with $V_{BS} = 0$ V, the Fermi levels inside the p-GaN and n-GaN layers are equal ($E_F = E_{F(n+)} = E_{F(p)}$). The formed inversion channel at positive V_{GS} can be populated freely from the source and drain regions (a) and c)). A negative body bias increases the potential of the bulk p-GaN and creates an additional barrier to the source and drain side (at low V_{DS}), which is shown in b) and d). Carriers from the source and drain have to overcome this barrier. This is only possible if the gate bias is further increased in order to reduce the barrier. The change of the surface potential for strong inversion becomes $\phi_S = 2\phi_F - V_{BS}$. Hence, the threshold voltage increases for negative V_{BS}. The effect of a body bias on the threshold voltage is already included in Equation (2.34). By reformulating (2.35) one obtains

$$N_{A-} = \frac{\gamma^2 C_{ox}^2}{2q\varepsilon_0\varepsilon_{r,GaN}}, \tag{2.39}$$

with which it is possible to estimate the ionized dopant concentration, as it is shown in our work in [83]. From (2.34) it can be seen, that the derivative of V_{th} by the p-GaN potential $\sqrt{2\phi_F - V_{BS}}$ equals γ. But, since ϕ_F depends also on N_{A-} the solution has to be found iteratively with Equation (2.40), where n_i is the intrinsic carrier concentration of GaN [75, 84].

$$\phi_F = \frac{kT}{q} \cdot ln\frac{N_A}{n_i} \tag{2.40}$$

Due to the pn junction formed between source and body it is only practical to apply a potential at the body, which is more negative related to the source. At positive bias the junction is biased in forward direction and conducts a higher current, which leads to high voltage drop within the highly resistive p-GaN layer and at the contact. Advantages of this method are the evaluation of the relative shift of V_{th}. Since the direct evaluation of N_{A-} from the absolute V_{th} is much more sensitive on trapping effects, in particular on Q_{fix}.

3 Fabrication and characterisation

In this chapter the fabrication and characterisation methodology will be explained. First the growth of the starting material stack is explained and an overview over different III/N growth methods is given. The influence of the substrate and layer properties such as dislocation density, doping density and layer thickness are crucial for later device operation. Considering this, the micro-structural methods for characterisation of the obtained layer stack after different growth steps are provided and discussed. Next, in Section 3.3 the device processing is discussed, separately for etching, formation of ohmic contacts and for the gate module fabrication. Possible variations of the processing scheme are introduced. In Section 3.4 special electrical characterisation structures are described.

3.1 Growth methods for GaN substrates and layers

In order to enable the complete potential of the GaN-material system while simultaneously keeping focus on the cost and growth effort it is important to utilize the most suitable growth process for a certain functional layer within the required vertical structure. Different CVD and PVD growth methods have been developed, which can be used for special growth steps and shall be explained shortly.

Hydride Vapour Phase Epitaxy (HVPE) is on of the most promising CVD methods to grow thick bulk GaN crystals. Within a vertical or horizontal arrangement a 800 °C liquid Ga reservoir is exposed to hydrocholoric acid (HCl). The thus formed $GaCl_3$ acts as metal precursor and is transported in a carrier gas flow consisting of nitrogen and hydrogen to the substrate. As nitrogen supplying specimen ammonia (NH_3) is used in conjunction with similar carrier gases. The rotating substrate containing a GaN seed layer is heated to around 1000 °C. GaN is formed with a rate of up to 100 µm/h by the reaction of both precursors at atmospheric pressure on the substrate surface. N-type doping can be established by a carrier gas flow around a solid dopand sources e.g. germanium (Ge) or addition of gaseous materials like dichlorsilane (DCS). [85, 86, 87]

A widely used growth method for GaN is **Metal Organic Vapour Phase Epitaxy** (MOVPE) due to its application in LED industry. Moderate growth rates and flexible precursor-based doping allow either well controlable growth of ternary and quarternary heterostructures as well as growth and compensation of several micrometer thick buffer layers. The process is based on a mid pressure (50 - 250 mbar) reaction in a temperature range of 950 - 1150 °C with trimethylgallium (TMGa) and ammonia (NH_3) for GaN. In combination with trimethylaluminium (TMAl) AlGaN can be grown. N- and p-type doping is accomplished by incorporation of silicon and magnesium, respectively. Typical precursors for that are silane (SiH_4) or bis(-cyclopentadienyl) magnesium (Cp_2Mg). Usual growth rates are around 2 µm/h [88, 89, 90].

Molecular Beam Epitaxy (MBE) is a typical ultra high vacuum (UHV) PVD method for the epitaxial growth of GaN at lower temperatures compared to HVPE and MOVPE.

On the one hand the low base pressure ($< 1{\cdot}10^{-9}$ mbar) of the growth environment is beneficial for high purity and hence very low impurity concentrations in the obtained layers. On the other hand the process is impractical for the growth of layers thicker than 2 µm due to the low growth rate. The growth temperature can be varied in the range of 650 - 800 °C and is usually kept below the decomposition temperature of GaN, which is around 800 °C under these conditions. During growth Ga and Al are supplied by evaporation from effusion cells. Nitrogen species can be supplied in two ways. In ammonia MBE nitrogen is supplied by ammonia as precursor, whereas in plasma assisted MBE (PAMBE) reactive nitrogen species are transported to the substrate after the decomposition of molecular nitrogen in a plasma cell [91]. This methods is used in this work. Doping in an MBE system can be established by an additional implementation of silicon to provide n-type doping or magnesium for p-type doping, based on evaporation or precursor addition. Common growth rates are in the range of some 100 nm/h. Further information can be found in [92] as overview and more specifically for the setup used here,[93].

3.2 Substrates and the desired starting material

In this section a closer look to the layer stack of the pseudo-vertical and true-vertical MOSFET and its characterisation structures will be given. Two approaches have been applied using either sapphire based substrates or bulk GaN substrates grown by MOVPE or HVPE, respectively. Further growth steps are necessary to obtain the finally desired $n^+/p/n^-/n^+$ layer stack as shown in Fig. 3.1. The properties of the functional layers are strongly impacting the final device properties. Finally, micro-structural characterisation methods are discussed. Special insight is given to dislocations and impurities typically present in different concentrations in the various layers and templates.

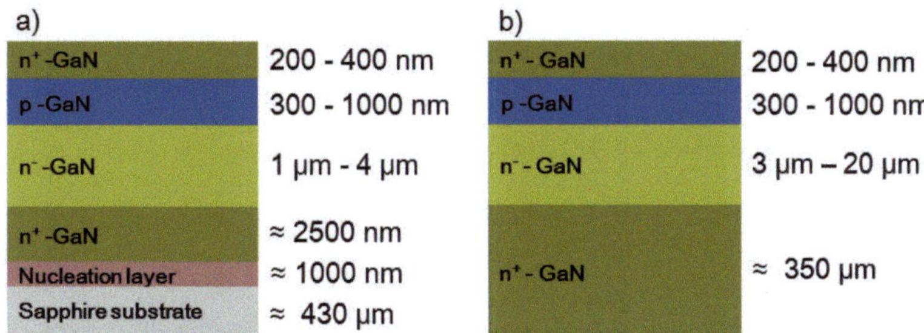

Fig. 3.1: Two different layer stacks obtained by the usage of sapphire based substrates (a)) or bulk GaN substrates (b))

Besides GaN on sapphire or bulk GaN substrates additionally GaN on Si and GaN on SiC substrates are available. In this work 2 inch GaN on sapphire templates were used to investigate key characteristics of the layers and devices. Their availability is good and they provide lower cost. Further investigations on 2 inch bulk GaN substrates have been performed with an application near device design and a thicker n⁻-GaN drift layer of about 4 µm.

Typical commercial substrates are specified by a certain c-plane orientation offset angle (offcut). In general a small offcut is beneficial to establish good nucleation during growth on the substrate. Typical offcut values after the substrate production are 0.1 - 0.5° [94]. A schematic illustration of the offcut angle is given in Fig. 3.2.

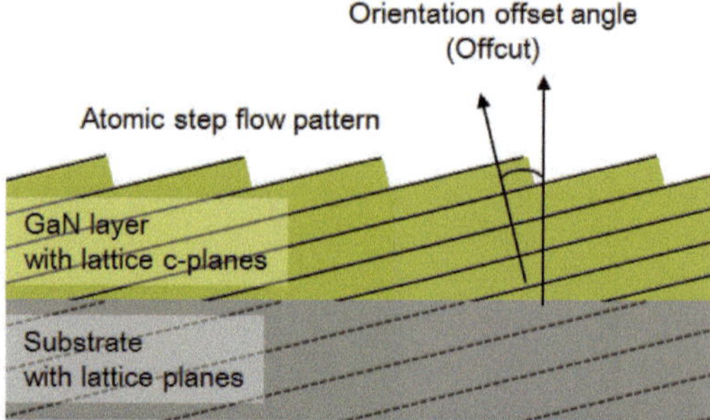

Fig. 3.2: Schematic cross section showing the orientation offset angle (offcut) between the c-plane normal vector of the GaN lattice and the normal of the surface, which represents the growth direction. The offcut is typically transferred from the used substrate with different or similar lattice parameters during epitaxy.

The offcut reduces the necessary diffusion length of adatoms on the surface by providing nucleation centers at the step edges. The formation of a more stable growth front along the atomic steps is provided. A deeper inside is presented in [95] and [96]. An atomic step flow pattern with similar offcut is created. In case of a planar surface, which is parallel aligned to the substrate the offcut is represented by the step height to terrace length ratio. In general, the surface morphology can be strongly influenced under large variations of the offcut. However, under small offcut angles between 0.1 and 0.5 ° no strong influence of the grown surface morphologies were found with respect to the defect density or roughness of the layers.

3.2.1 Physical and micro-structural characterisation

The surface morphology of GaN can be investigated by **Atomic Force Microscopy** (AFM). With an atomically sharp tip placed on a bendable cantilever the sample surface is scanned and the deflection of the cantilever is measured by a laser. A detailed explanation of the working principle can be found elsewhere [97]. Combining the information of all three spacial directions a three-dimensional colour-coded map of the surface can be obtained and profiles with the height information can be extracted. In this work a *DI/Veeco Dimensions 3100* AFM was used in tapping mode to investigate the sample surfaces. For the data analysis, imaging and profile as well as roughness extraction the open source software *Gwyddion 2.3.4* was used.

Scanning Electron Microscopy (SEM) is a suitable method to analyse chemical and structural properties of the sample. Inside a high vacuum chamber an electron beam is scanned over the surface of the sample. With a secondary electron (SE) detector scattered electrons from surface near regions are detected. The signal is plottet versus the position of the electron beam and thus a 2D scanning electron micrograph is obtained. Contrast is created by sample edges, different materials or doping concentrations and at differing surface inclinations. **Transmission Electron Microscopy** (TEM) is very similar but differs in the electron energy (50 keV - 500 keV for TEM compared to 1 keV - 20 keV in SEM), increased spatial resolution and the sample geometry. Whereas in SEM typically bulky samples are monitored in a reflective way, in TEM nanometer thin sample slices have to be prepared for a transmissive investigation. A detailed explanation of both methods can be found in [98, 99]. In this work a *Zeiss Leo 1560* was used for SEM and TEM analyses were performed with a *Zeiss Libra 200 HR*.

Elemental depth profiles can be obtained by **Secondary ion mass spectrometry** (SIMS). In this destructive method the sample surface is exposed to a focused caesium ion (Cs^+) beam with an ion energy of 14.5 keV. It leads to sputtering of surface near sample atoms. Charged species are selectively counted by passing into a mass filter and spectrometer. From a reference sample with known elemental concentration the unknown element concentration in the sample under test can be calculated. Due to sputter etching and removal of material from the surface deeper layers are consecutively analysed with a depth resolution of around 10 nm. In this work a *Cameca IMS 6F* tool is used to obtain the depth distribution of Si, Mg, C and O in the doped GaN layers. Further information about SIMS on GaN can be found e.g. in [100].

3.2.2 Dislocations and impurities

Dislocations are one dimensional defect structures formed during growth on discontinuities of the crystal structure such as phase transitions, grain boundaries or highly disruptive zero-dimensional defects or clusters. The density of dislocations is obviously related to the amount of crystal imperfections introduced by these defects. Forming along the growth direction threading dislocations can run through the complete layer stack or substrate. Without special provisions, continuous GaN growth leads to a mutual annihilation of dislocations by one order of magnitude for each increase of the overall grown layer thickness by a factor of 10 [101]. A low dislocation density in overgrown material can be obtained by using substrates with almost no lattice mismatch and inherently low defect density e.g. preferably bulk GaN substrates or by very thick GaN growth on foreign substrates. The later can principally be used to obtain freestanding GaN substrates by separation of the underlying substrate, but is not preferred for the vertical device technology on foreign material. Thick layers on foreign substrates cause high stress within the GaN stack and a sophisticated stress management structure has to be applied. In this way a dislocation density of smaller than $1{\cdot}10^8$ cm^{-2} can hardly be achieved and a higher density is typically persistent [102] for reasonable growth times.

Recently, several reports were published to understand the nature of dislocations often separated in three basic types, which are threading edge dislocations (TED), threading screw dislocations (TSD) and mixed type dislocations (TMD) [103, 104, 105]. Larger defects can be observed as cracks or nanopipes originating from the nucleation or induced by stress inside the layer stack and formed by the superposition of threading dislocations [106].

Especially TSD and TMD terminating on the surface of the grown layers can be easily investigated by AFM, due to the formation of larger surficial defects at their termination points with a diameter up to several nanometres. Further, an additional terrace step begins at their termination point. Often the densities of TSD and TMD are summarized because a clear discrimination by AFM is impossible. Decoration by etching or during the growth is possible and further discussed in Section 3.3.2 and Section 4.6. However, for subsequent processing these methods are not preferred. In contrary, TED termination points can be located on a terrace and the decoration is less pronounced under usual growth conditions. Thus, the evaluation of TED is more difficult and requires sufficiently high AFM resolution and surface quality. The latter is not always given.

Figure 3.3 presents two AFM images of the surface morphology of GaN layers grown by MOVPE and MBE. A clear discrimination between different defects can be made on these micrographs. Especially, in the case of the MBE grown GaN layer deep pits with a hexagonal shape are formed on TSD and TMD, whereas TED form smaller and shallower pits. Mostly they align along lines, which are likely formed by small-angle grain boundaries [107, 108].

From an impurity point of view a discrimination between the different growth methods is done. In an UHV ultra-pure MBE system a background impurity concentration of $<$ $2 \cdot 10^{16}$ cm^{-3} can be reached. The main residual species is oxygen. Its concentration depends strongly on the growth temperature [93]. Since MBE is utilized in this work for the growth of the source layer with higher n-type doping the residual n-type doping with a level below $5 \cdot 10^{16}$ cm^{-3} is not critical.

For MOVPE growth typically hydrogen and carbon background dominates down to a concentration of $1 \cdot 10^{16}$ cm^{-3}. This is, in particular important for the drift layer doping. A lower UID level provides the possibility of low intentional n-type doping concentrations. Further, carbon is reported to act as deep trap and compensation agent in GaN [109]. A low concentration is thus beneficial concerning trapping in the bulk drift layer.

The here used HVPE-layers serve as highly doped n^{+}-GaN substrates and are thus neither critical in terms of impurity concentration nor from a trapping point of view.

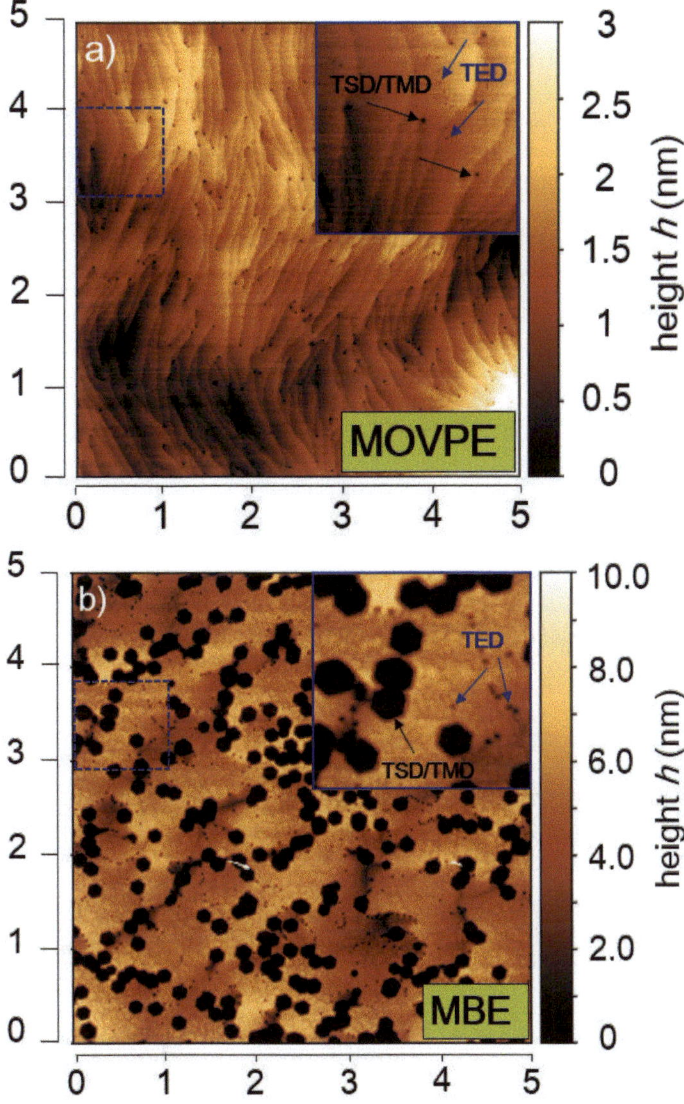

Fig. 3.3: AFM images of an area of 25 µm^2 from two differently grown GaN layers by MOVPE (a) and MBE (b). A typical surface morphology with a bimodal dislocation decoration can be observed. Small pits with a density of around $2 \cdot 10^9$ cm^{-2} are associated to TED, whereas the larger pits with a density of around $1 \cdot 10^9$ cm^{-2} are likely formed around TSD and TMD, respectively.

3.3 Pseudo- and true-vertical MOSFET fabrication

In this section the fabrication of true-vertical and pseudo-vertical MOSFETs is presented. The complete processing of the devices in this work is based on contact lithography on two inch GaN templates either based on sapphire or bulk GaN. The breakdown voltage of (pseudo-)vertical GaN structures such as pn diodes and FETs can be strongly dependent on the used substrate and base material. In particular, their dislocation density and impurity concentration can cause a significant influence on the breakdown and leakage behaviour as well as on the contact formation or the dielectric interface. Additional characterisation structures and methods are very helpful to monitor effects stemming from different templates to verify their significance.

3.3.1 Processing routes

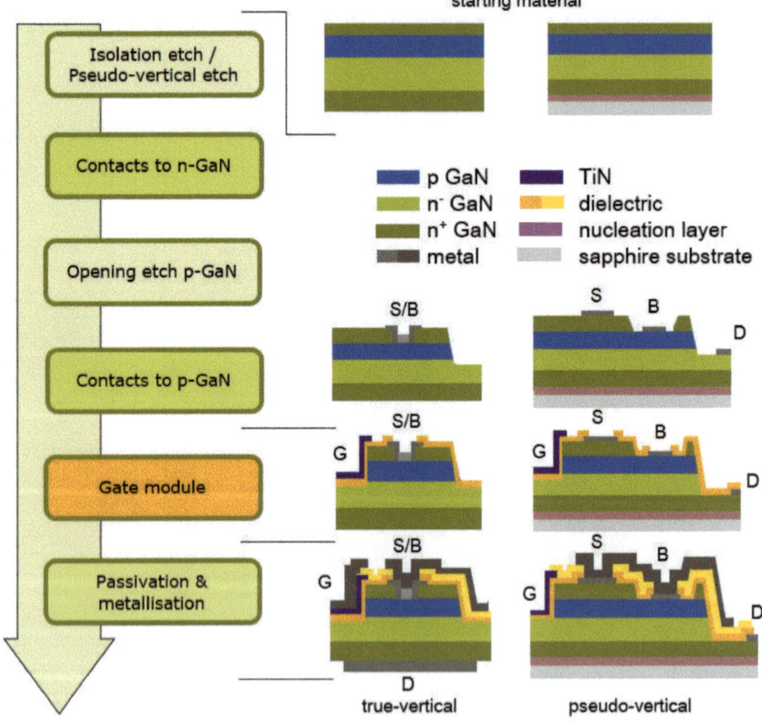

Fig. 3.4: Schematic illustration of the process flow with the respective schematic device cross sections for a true- and pseudo-vertical MOSFET. The cross sectional view of the devices are shown after different processing blocks, which are the isolation, p-GaN contact formation, gate module fabrication and metallisation as shown by the black lines. Gate, Body, Source and Drain are denoted by G, B, S and D, respectively.

The base process has 8 lithography steps, which are explained in the following. The pseudo-vertical approach based on GaN on sapphire templates was first used to evaluate the different process steps and their interactions. Next the processes were transferred to a true-vertical device utilizing a bulk GaN substrate. Restrictions due to the formation of the drain contact on the wafer top-side for the pseudo-vertical approach are introduced already in Section 2.2, but will be addressed again for completeness. A schematic of the different device structures after the respective process modules is shown in Fig. 3.4. The different steps are closer explained in the following, whereas this description only refers to the basic processing scheme. Modifications or re-ordering of the process blocks is possible, in particular for the steps before the gate module.

1. **Isolation etching / pseudo-vertical drain etching**

   The Isolation etching / pseudo-vertical drain etching acts as device isolation for the true-vertical MOSFET as well as for the pseudo-vertical MOSFET. Additionally, the etch has to access the n^+-GaN drain layer for the subsequent contact deposition in the pseudo-vertical device. From the isolation perspective, it is sufficient to etch through the p-GaN in order to separate the source and body for adjacent devices. In this respect, the necessary depth is around 1.3 µm related to the combined thickness of the source (200 nm - 400 nm) and body layer (300 nm - 1000 nm). For deposition of the drain contact in the pseudo-vertical device the drift layer has to be etched in order to obtain low contact resistances. For deep etching a mask material is needed which can resist the long etching times. In this work an AZ5214E image reversal resist mask with a negative polarity and a BCl_3/Cl_2 inductively coupled plasma (ICP) dry etching step is used. The resist/GaN selectivity of the etching recipe is around unity. The maximum etch depth is thus limited to depth of around 3 µm, for a resist mask with a thickness of 4 - 5 µm. The drift layer thickness should additionally not exceed 1.5 µm due to additional lateral contraction of the resist mask, which causes a flattening of the edge with increasing etching times.

2. **Ohmic n-GaN contact formation**

   A contact metal stack consisting of 20 nm titanium, 100 nm aluminium, 40 nm nickel and 100 nm gold is deposited by electron beam evaporation after preparation of a AZ5214E lift-off resist mask (positive polarity). After the deposition and subsequent lift-off using acetone the samples underwent a rapid thermal anneal (RTA) in nitrogen for 30 s at different temperatures. For all RTA processes described in this work an *AST SHS 2800* RTP system was used. The utilization of a hard mask is avoided, due to the increased process complexity.

3. **p-GaN opening**

   In a similar process as described for the isolation/pseudo-vertical etch the p-GaN is accessed. The comparably shallow etch has a depth of around 110 % of the source layer thickness stopping inside the p-GaN.

4. **p-GaN contact formation**

   The p-GaN contacts were formed in two different approaches. The first, includes nickel and subsequent gold evaporation with a thickness of 20 nm for both layers. The contacts are structured by lift-off and afterwards annealed in an RTA process in N_2/O_2 mixture with 25% oxygen at 550 °C for 5 min.

The second approach includes the deposition of titanium nitride (TiN) [110] by reactive sputtering from a Ti target with an Ar/N_2 plasma and subsequent structuring by wet chemical etching with standard clean 1 (SC1) containing ammonium hydroxide and hydrogen peroxide followed by an RTA at 700 °C for 30 s in pure nitrogen.

5. **Gate module processing**

Whereas the previously described process steps involve only one lithography step a dual lithography module is described in the following. First the **gate trench formation** is performed by etching with an BCl_3/Cl_2 ICP etch through the n^+-GaN source and p-GaN body layer into the n^--GaN drift layer to ensure a good channel connection to the drain. This etching step is optimized for selectivity to the mask material and smooth as well as steep edges. A plasma enhanced chemical vapour deposited PECVD Si_3N_4 is used as hard mask, which can be covered by Ni in order to improve the selectivity of the etching recipe. The hard mask, either with or without Ni cap, is deposited and structured before the actual trench formation by a resist mask. Structuring is done using an SF_6 based ICP etch prior to the gate trench formation for the Si_3N_4 and by lift-off for the Ni. After the reactive ion etching a wet chemical etch at 80 °C with tetramethylammonium hydroxide (TMAH) for one hour is performed. It removes the plasma damaged surficial layers and steepens the gate trench edge in order to improve the later channel properties and trapping characteristics. At last, the hard mask is removed. The Ni can be etched by diluted HNO_3 to around 10 %. The ohmic contacts are protected by the nitride hard mask during this step. Afterwards the Si_3N_4 is removed with 5 % hydrofluoric acid, which also pre-treats the vertical channel surface.

After the formation of the trench the **gate stack processing** is started. A thermal ALD Al_2O_3 deposited at 300 °C using trimethylaluminium (TMAl) and ozone (O_3) serves as gate dielectric. The used ozone process results in optimized leakage behaviour of the oxide. In order to improve the interface to the GaN channel the first ALD cylces can be performed with water as oxygen precursor, the process is described closer in [78]. Subsequently a 10 nm ALD TiN gate metal layer is deposited with a titanium(IV)-chloride ($TiCl_4$)/NH_3 based process at 250 °C. The layer thickness is increased to 50 nm by reactive sputtering with the previously described process for the p-GaN contact to lower the sheet resistance of the gate metal stack. The gate metal is pre-structured with a resist mask and an ICP etch in chlorine atmosphere. To finalize the structuring the same resist and a SC1 etch is used, which removes the remaining TiN on the unmasked area. The previous thinning by the ICP etch prevents an early lift-off of the resist mask, due to the otherwise longer exposure to the SC1 solution.

6. **Passivation and Metallisation**

In general the intrinsic pseudo-vertical device can be finished after the gate module by opening the passivation on the metal pad area with a wet chemical HF etch. For true-vertical devices a back-sided metallisation has to be deposited in order to form the drain contact on the backside of the wafer. Additionally the substrate can be thinned in order to reduce its resistance contribution to the ON-resistance of the device. Caused by the higher target OFF-state operation voltage of true-vertical devices a field plate at the lateral device termination is considered. Further dielectric layers such as Si_3N_4 or SiO_2 can be deposited followed by a resist based opening of the contacts and an evaporated metallisation layer forming the field plate and contact interconnections. Structuring can be performed by metal lift-off. A wafer obtained before further passivation and higher metallisation is shown in Fig. 3.5.

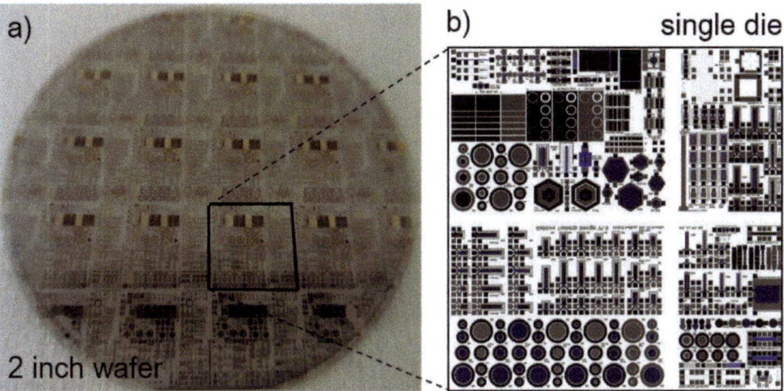

Fig. 3.5: Picture of a completely processed two inch wafer (a) and zoomed top view of the layout of a single die, which was processed on the wafer (b).

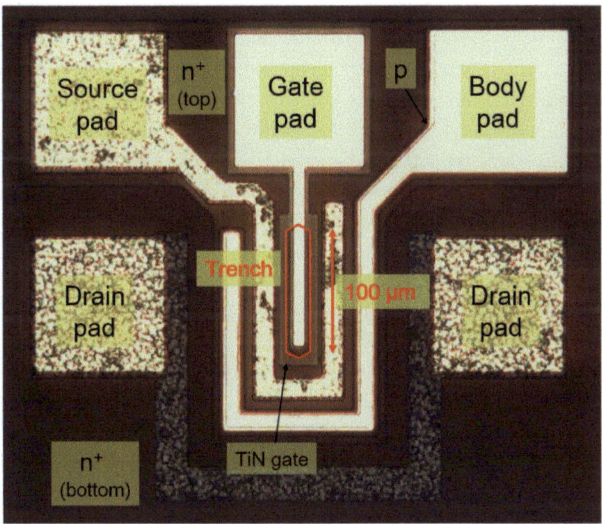

Fig. 3.6: Optical photomicrograph in bright field (BF) of a pseudo-vertical MOSFET after the processing of the upper metallization layer with an overall gate width of 200 µm. The trench of the device is marked by the red hexagon along the trench edge.

A top view of a pseudo-vertical device with a linear trench of 100 µm is shown in Fig. 3.6. The channel width is 200 µm. The device is completely processed with higher passivation of 40 nm ALD Al_2O_3 and Ti/Al/Ti/Au metallisation. The pad structure is build up by the contact metallisation layers and additional upper metallisation located on the respective layers which are contacted. Due to the annealing of the n-GaN contacts the layers are rough. This can be seen in Fig. 3.6 by less reflection of the respective pad and contact structures. The device utilizes no field plate. However, various designs of the MOSFET are included in the chip design. So it is possible to address and validate the influences of different design parameters and modifications.

3.3.2 Inductively-coupled plasma etching

Reactive ion etching (RIE) plays an essential role in the processing of the pseudo- and true-vertical devices, since it enables to partially remove functional layers to access others in a dry chemical way. Inductively coupled plasma (ICP) RIE is an advanced method to independently control the ion concentration and ion energy during the process by implementation of an additional inductively coupled power transmission to the plasma. A *Plasmalab System133* from *Oxford Instruments* with an *ICP380* source is used in this work. While the ion density can be controlled by this ICP source, the energy of the ions is controlled by a capacitvely coupled susceptor electrode similar to conventional RIE. By this principle the etching of different masking schemes, etch depth and etching profiles can be tuned in a more flexible way by independent control of physical i.e. sputter etching, and chemical etching. Further the recipes can be optimized in terms of surface morphology, edge slopes and selectivities to the respective masks. Important parameters include the power, temperature, pressure and chemistry during the etching. The interplay between these parameters was investigated by various previously published works like [111, 112, 113]. Figure 3.7 represents the changes of the surface morphology, which can be achieved by changing only one of the called parameters. Here the RF power of the ICP etching process was varied. At 300 W RF-power the surface features bubble like structures and is unfavourable for further processing. Lowering of the RF-power towards 150 W reduces the amount of residual surface structures till a smooth surface and a steep etch flank is obtained. Similar dependencies can be found for the decoration of dislocation related defects on the surface. Generally, stronger chemical etching by reactive species within the etching atmosphere leads to a stronger decoration of defects (e.g. at higher chlorine content) and shallows the edges of the trench. A higher physical etching contribution instead leads to stronger sputtering, more anisotropic etching but also less selectivity to other materials and stronger plasma damaging. Zhang et al. investigated these effects related to the trench formation for vertical GaN MOSFETs [114].

Typically plasma etching converts exposed p-GaN surface to n-type GaN which e.g. strongly degrades contacts to the layers. One interesting approach to achieve low plasma damaged surfaces was shown in [115] with a digital etch as gate recess in AlGaN/GaN HEMTs. Similar processes including an alternating O_2 and BCl_3 plasma etch with low RF power seem possible for pure GaN in order to obtain low damaged surfaces [116].

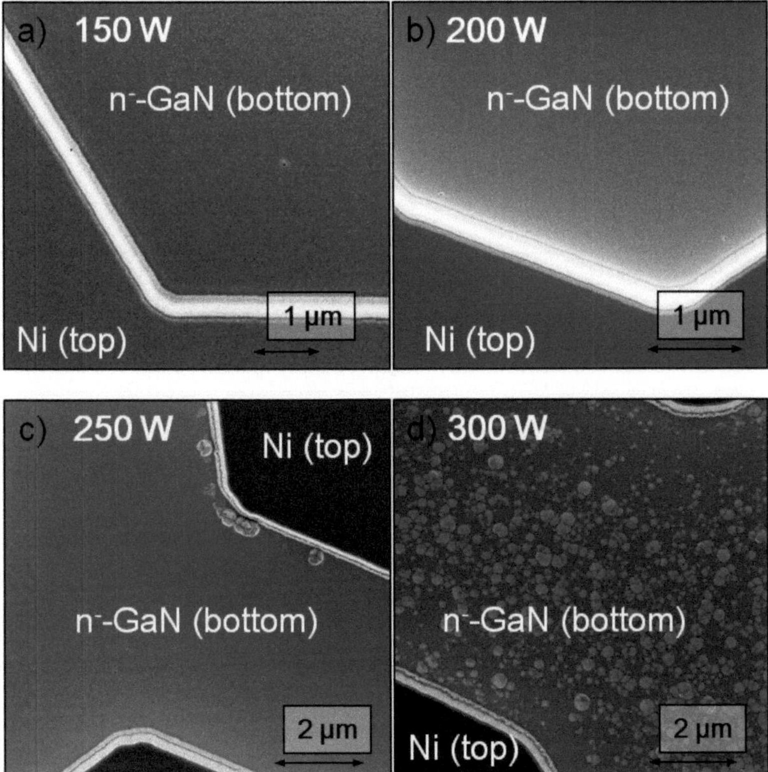

Fig. 3.7: Inclined top view of the gate trench obtained by SEM, etched under different RF-power conditions. Other process parameters where kept constant.

In contrary to the shallow etch, for accessing the p-GaN body layer, the isolation etch has to reach the n⁻-GaN for true-vertical designs or even the n⁺-GaN layer for pseudo-vertical stacks. Thus a higher etch rate is more practical, which in turn produces increased plasma damage. One possible trade-off is the application of both, first a high rate process and subsequently a low damage low rate process. In case decoration of defects occurs it has to be considered, that for pseudo-vertical stacks the dislocation density rises with increasing etching depth. The trench used for the gate can be etched with higher rate since subsequently a wet chemical treatment is implemented, in order to remove the damaged layer. Here, the selectivity to the hard mask and the steepness of the trench edges are focused.

3.3.3 Process flow modification

In this section a possible variation of the process flow is presented. Both approaches provide advantages and disadvantages and the application of either flow has to be evaluated with respect to the subsequent processing of the gate module, passivation and metallisation.

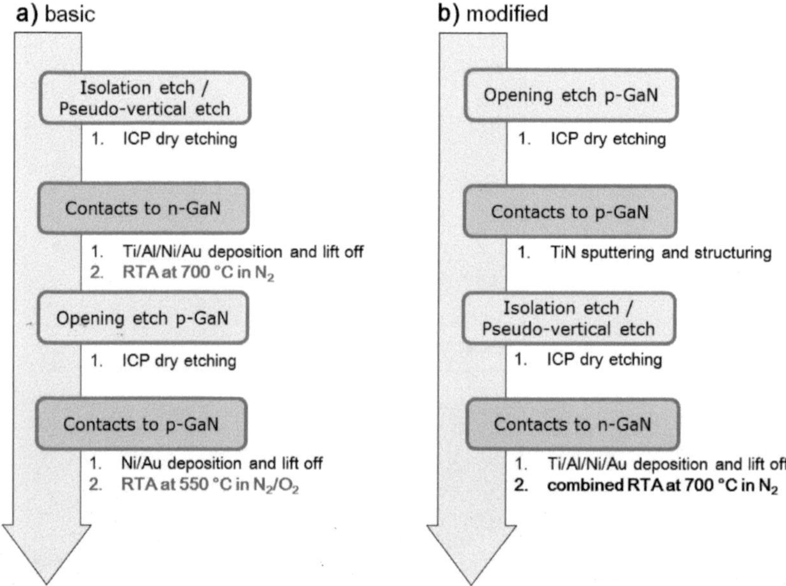

Fig. 3.8: Schematic, simplified first part of the process flow. In the flow chart a) the basic processing scheme is shown with alternating etching and contact formation. In b) the modified processing scheme using TiN p-GaN contacts and a single RTA in a changed sequence is shown.

In Fig. 3.8 a) the basic processing flow before the gate module using Ni/Au based p-GaN contacts is shown. With the separation of the processing of the two different metal contacts it is possible to protect the p-GaN surface from the RTA performed for the contacts to n-GaN. One drawback in this scheme is the higher thermal budget of the two annealing steps. Additionally in this sequence the ohmic n-GaN contact is annealed first in its dedicated RTA step and later again in the RTA step of the p-GaN contact under O_2 presence. Oxidation of the metal layers can thus occur. The alternative process allows the exclusion of oxidation during subsequent anneals.

The modified process flow (Fig. 3.8 b)) starts with an n^+-GaN etching step to access the p-GaN. Next, the TiN p-GaN contact is deposited and structured. Afterwards the isolation/drain layer opening and second the n-GaN contact formation is made. Finally, the contacts are annealed in a single RTA under pure nitrogen atmosphere without oxygen addition. The TiN p-GaN contact plays an essential role in this scheme since its RTA can be similar to the n-contact RTA reducing the overall thermal budget of the process flow. The chemical resistance against wet treatment is drastically reduced for non-annealed Ti/Al/Ni/Au contact stacks. In particular, this prohibits a strong chemical treatment of the exposed contacts after the contact deposition and before the RTA. The better chemical stability of the TiN is beneficial in later processing and concerning pretreatment of the p-GaN surface. But as described in Section 5.2.2 the contact resistance to the p-GaN is increased compared to the Ni/Au contact stack with RTA at 550 °C in oxygen containing atmosphere.

3.4 Electrical characterisation, structures and process control

In this section electrical methods to characterise the fabricated device are explained, sources of potential errors are discussed and the related mathematical functions are introduced. Beside quasi-static or static measurement methods like breakdown measurements, $I(V)$ and $C(V)$ measurements different characterisation structures are discussed. The used wafer and die design features characterisation structures, which can be used to investigate different MOSFET modules separated from the actual MOSFET. As example pn diode structures, MIS-structures and TLM structures are shown. In order to screen different process steps of the previously described process flow it can be necessary to perform accompanied measurements on characterisation structures in an intermediate step of the process flow. Thus interactions between different process modules and steps can be figured out. For $I(V)$ and $C(V)$ measurements a *Agilent B1505A Power Device Analyser* is used in conjunction with a *Karl Süss PM8 probe station* and a 200 mm thermo-chuck from *att systems*.

3.4.1 Current voltage characterisation

For static or quasi-static $I(V)$ measurements a bias or current is applied to the device terminals. The bias or current is changed stepwise and both, current and voltage are simultaneously measured on each step after settling of dynamic signal contributions. In order to characterise the devices dynamically pulsed $I(V)$ measurements can be performed, which allow to reduce the device self heating. They target rather fast dynamic responses of the device.

Under consideration of the geometrical dimensions on various test structures conclusions about sheet and layer resistances, leakage current and threshold or breakdown voltages can be obtained. In combination with $C(V)$ measurements, which are described in the following section, additionally carrier concentration and mobility estimations are possible. Special $I(V)$ characterization structures are presented in Section 3.4.3.

3.4.2 $C(V)$ measurements and charge carrier profiling

During the static capacitance-voltage measurement a step-wise DC-bias voltage superimposed by a small signal AC-voltage is applied between two terminals of the device. The AC-current is measured for each DC-bias step. With the assumption of an equivalent circuit the complex admittance between the terminals can be calculated dependent on the bias voltage. If a reactance is present the current through the DUT will be phase shifted. In case of a pure capacitive behaviour the current leads in phase by 90° and from the amplitude the size of the capacitance can be obtained. The DUT shows non-dissipative behaviour in this case. If a mixed/complex admittance is present the DUT shows dissipative behaviour and the calculation will result in a resistive and reactive component according to the equivalent circuit assumed. Since, from voltage and current only two values can be measured (amplitude and phase shift) only a two-component equivalent circuit can completely be described. As long as the dissipative component within the circuit is small, simple equivalent circuits like a series or parallel arrangement of resistor and capacitor will yield the same capacitance. If a higher dissipation occurs two cases are possible.

- the dissipative element can be approximated by one of the two previously mentioned models; measurements at different frequency can help to discriminate
- non of the two models is adequate and without additional information no valuable result can be obtained

As guidance a dissipation factor of < 0.1, which correlates to a phase shift of $5.7°$ from pure capacitive behaviour, can be used to consider the dissipation as negligible.

From reliable measurement values a charge profile can be obtained, where the dielectric thickness and charge concentration according to a parallel plate capacitor model can be calculated by Equation (3.1) and (3.2). If one electrode is spatially fixed e.g. is formed by a highly conductive layer and the dielectric is formed by a space charge region within a semiconductor, the depletion depth can be calculated by:

$$z = \frac{\varepsilon_0 \varepsilon_r A}{C(V)}.$$ (3.1)

By calculation of the charge concentration a depth profile can be obtained. In contrary to the depth calculation, the apparent charge density is not only dependent on the absolute capacitance but also on its voltage derivative according to:

$$N = \frac{C(V)^3}{q\varepsilon_0 \varepsilon_r A^2} \frac{dV}{dC(V)}.$$ (3.2)

Finally in this section, variations of the extracted doping and carrier concentration shall be discussed shortly with use of the screening length

$$L_D = \sqrt{\frac{kT\varepsilon_0\varepsilon_r}{q^2(p+n)}}.$$ (3.3)

which is also called the Debye length L_D. It describes a characteristic length or depth for the change of the potential landscape inside the material. It thus reflects the distance at which irregularities from the equilibrium drop by a factor of $1/e$ (to $\approx 37\%$) and hence it describes the spatial extension for changes of the potential landscape and the adaption of the carrier concentration. Equation (3.3) describes the calculation of the Debye length in a general way.

If GaN is doped, one kind of carriers (p or n) can be neglected, due the wide band gap and low intrinsic carrier concentration. It can be seen, that for low doping concentrations the Debye length has to be taken into account when designing test structures for charge profiling. The layer thickness has to be larger than around $3L_D$ ($\approx 95\ \%$ drop) in order to obtain reliable information of a constant doping concentration [75].

For example for a carrier concentration of $1\cdot10^{16}$ cm^{-3} a layer thickness > 100 nm has to be considered on a low doped substrate. For even lower doping concentrations the charge profiling method gets impractical and thicker layers are needed. Further, a highly doped substrate is beneficial to provide a suitable backside contact. In this case the potential screening according to L_D occurs from both sides, from the metal electrode and from the highly doped layer below.

3.4.3 Cooperative characterisation structures

Besides the characterisation of the complete device in a $I(V)$ or $C(V)$ manner, different structures can be obtained during the processing. They can be used to analyse different aspects of the MOSFET device. The most important structures shall be closer explained in this section. Fig. 3.9 presents a circular MIS-structure used to characterise the dielectric layer, dielectric/n$^+$-GaN interface and the n$^+$-GaN layer. With the processing of similar structures on the p-GaN, n$^-$-GaN and lower n$^+$-GaN properties of all layers can be accessed. Therefore a MIS-stack has to be deposited surrounded by an ohmic contact to the same layer.

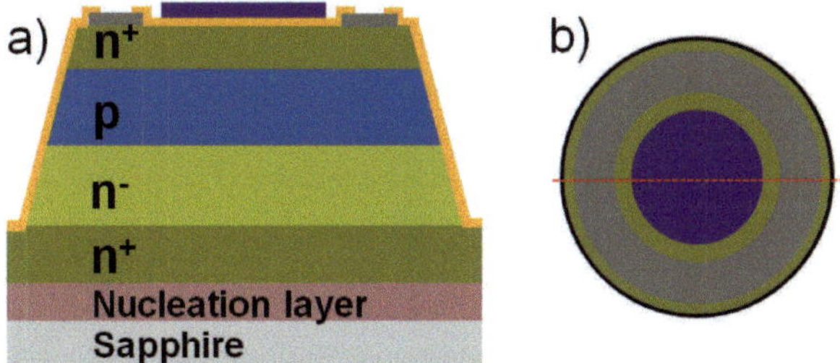

Fig. 3.9: Schematic cross-section (a) and top-view (b) of a circular MIS-structure on the source layer of a pseudo-vertical template. It is typically used to analyse the doping concentration and properties of the dielectric after processing of the gate layer stack.

On the one hand $I(V)$ characterisation allows the estimation of the breakdown and leakage behaviour of the dielectric, which is especially important for the gate trench bottom surface and the field plate. On the other hand, $C(V)$ measurements can be used to obtain the oxide capacitance and carrier density in the GaN layer. Trapping and interface related effects can be addressed by double or more dynamic voltage sweeping.

A circular pn junction test structure is shown in Fig. 3.10 with two ohmic contact layers on the adjacent layers. This structure can be used to separately characterise the pn$^-$ body diode of a MOSFET. In case a true-vertical stack is used, the lower n$^+$ layer can be contacted by a back-sided contact. By $I(V)$ measurements the reverse and forward characteristics of the body diode can be investigated. in particular, from the forward operation series resistance contributions are derived. The reverse leakage characteristics define the breakdown voltage of the pn$^-$ junction within the MOSFET. By the implementation of structures with different geometry and its influence on the leakage current, contributions scaling with the perimeter or area can be determined.

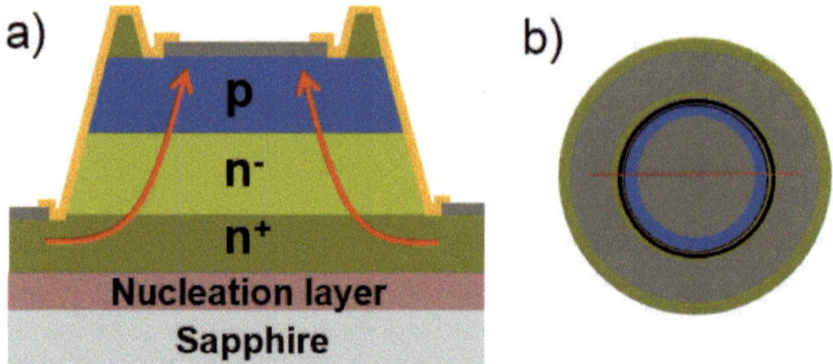

Fig. 3.10: Schematic cross-section (a) and top-view (b) of a circular pn⁻n⁺-junction with ohmic metal contacts on p- and n⁺-GaN of a pseudo-vertical template typically used to evaluate the body diode characteristics and contact properties.

A typically used linear TLM-structure for the estimation of the ohmic contact behaviour, the contact resistance to the different layers and the layer sheet resistances is illustrated in Fig. 3.11 with a cross-section (a) and top-view (b). Its evaluation is based on the $I(V)$ measurement of adjacent ohmic contact stripes on the same layer. Under the measurement conditions the electric field inside the layer should be small in such way, that the variation of the carrier concentration along the structure is negligible (e.g. below 10 kV/cm).

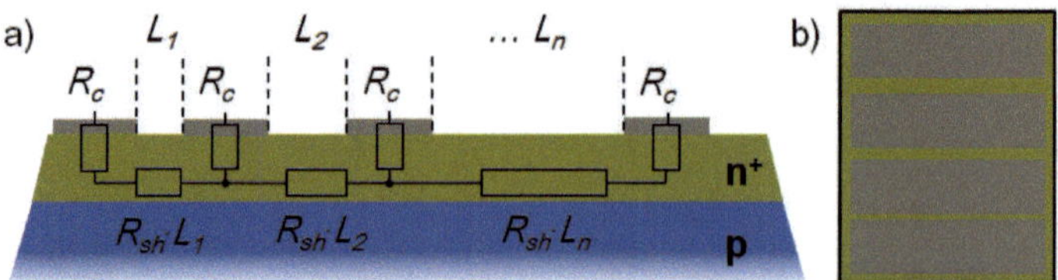

Fig. 3.11: Cross-section (a) and top-view (b) of a linear TLM structure on the source layer in order to evaluate sheet and contact resistance of the layer and contact material, respectively.

The resistance between two measurement probes (R_{TLM}) on two adjacent contact pads scaled to the width is given by

$$R_{TLM} = 2R_c + R_{sh}L \tag{3.4}$$

with the contact resistance R_c, whereas the layer contribution is dependent on the distance between the two contacts L, which is varied within the structure. From a plot of the measured R_{TLM} versus L the contact resistance and sheet resistance can be separately obtained. Uncertainties can arise from insufficient ohmic contact behaviour or from deficiencies in the geometry.

Two different options for the structuring of the contacts are possible. The first relies on a shadow mask process, where the TLM structures are deposited without application of lithography. Due to the resolution limits defined by the mask fabrication and by underdeposition at the edge of mask structures during evaporation and sputtering this method can only be applied for relaxed geometries. Furthermore, the semiconductor is usually not etched at the boundaries of the structure. The geometrical assumptions of a defined edge or isolation is thus not given. However, it can be approximated for large W/L ratios. In this work, large area structures with a width of 1 cm and L of 100 - 1000 µm are used. Although the area consumption is large for this processing method uncertainties arising from chemical surface treatments and resist deposition on the as-grown GaN surface can be avoided. Another benefit is given by the uncomplicated process sequence enabling quick feedback on split experiments.

The second option relies on the use of a lithographical approach, which includes wet chemical treatments for resist removal and development. The obtained structures are typically by a factor of 10 - 100 smaller with L ranging from 5 to 35 µm and a submillimeter width. The distance differences have to be kept in mind when applying a bias to the structures in order to keep the electric field inside the layers small. Additionally the structure is patterned in a defined lateral geometry and can easily be modified by subsequent lithography steps.

4 Properties of the functional layers

In Section 3.2 the required layer stack and common growth methods were described. In this chapter the characterisation data of the grown layer stack is shown and discussed. First, the growth sequence and the doping of the functional layers is presented. Next, in Section 4.1 and Section 4.3 the morphological features of the layers are described. Two approaches depending on the processed devices (i.e. pseudo- or true-vertical) are used in this work, which are shown in fig. 4.1.

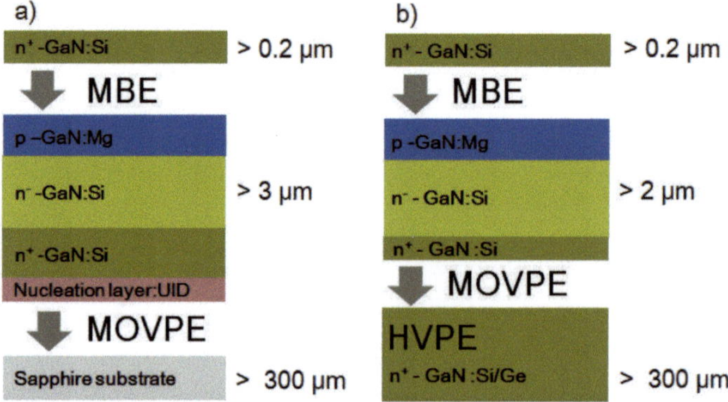

Fig. 4.1: Growth sequence of a pseudo-vertical layer stack on a sapphire template (a) and a true-vertical layer stack on a HVPE GaN template (b); note: before the MBE step a hydrogen out-diffusion anneal is performed.

Since the pseudo-vertical device is based on a sapphire template first a III/N layer has to be nucleated, which is usually AlN. Typically a dislocation density of around $1 \cdot 10^{10}$ cm^{-2} to $1 \cdot 10^{11}$ cm^{-2} is created at the template/nucleation layer interface [117]. A deep investigation on the nucleation was done by Koleske et al. [118]. After the growth of the drain, drift and body layer the dislocation density reduces by one to two orders of magnitude depending slightly on the thickness of the layers. After the MOVPE the dislocation density in the upper layers is thus typically in the range of $5 \cdot 10^8$ cm^{-2} - $3 \cdot 10^9$ cm^{-2}.

In case of the true-vertical MOSFET, as shown in fig. 4.1 b), a doped free-standing GaN wafer is used, which can vary in the dislocation density in the range of $5 \cdot 10^4$ cm^{-2} for ammonothermally grown GaN to $5 \cdot 10^7$ cm^{-2} for rather thin HVPE crystals. If a several millimetre thick HVPE crystal was used to obtain the bulk GaN wafer typically dislocation densities around $2 \cdot 10^6$ cm^{-2} are obtained. Due to the already strongly reduced dislocation density of the bulk substrate a further reduction during the growth of the few micrometer thick functional layers by MOVPE and MBE is not expected. In this work bulk GaN substrates with a dislocation density of $\leq 5 \cdot 10^6$ cm^{-2} were utilized.

4.1 Morphology of the MOVPE grown layers

In this section the surface morphology of the MOVPE grown layers with special focus on dislocation related defects and their decoration is discussed. Further, sapphire and bulk GaN substrate based layer stacks are compared.

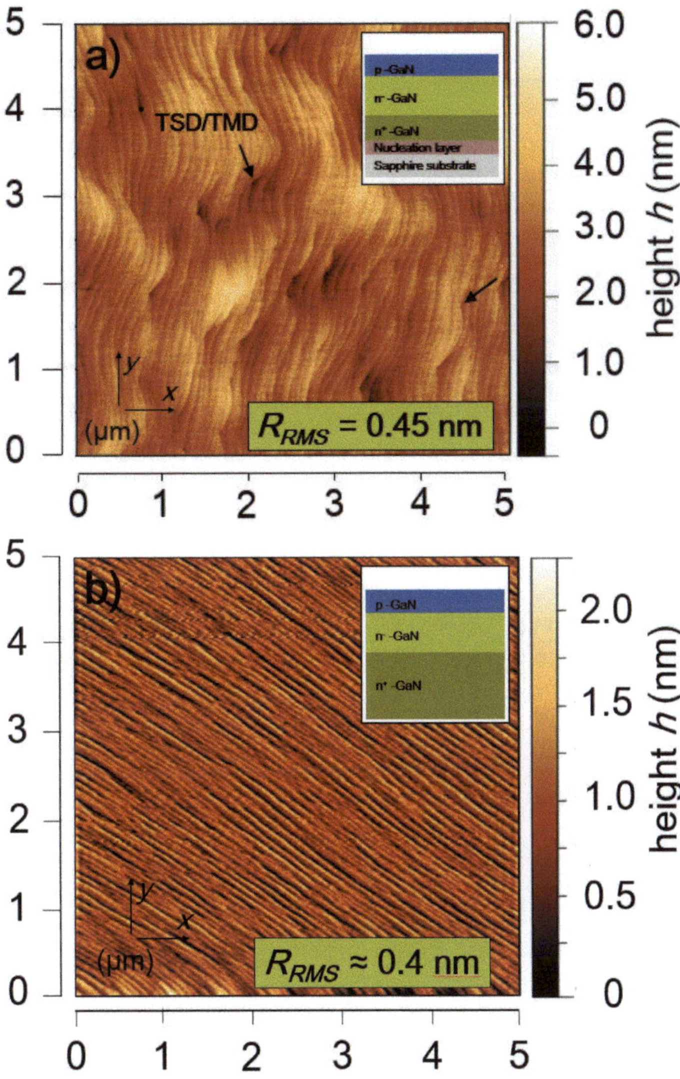

Fig. 4.2: AFM image showing the surface morphology of a typical sapphire based template obtained after the MOVPE overgrowth (a) and surface morphology of the MOVPE layer on a HVPE-grown bulk GaN substrate (b); both micrographs are obtained before the hydrogen out-diffusion treatment.

In fig. 4.2 a) the surface morphology investigated by AFM on a typical sapphire based template overgrown with MOVPE is shown. The overgrowth ended after the p-GaN with an overall smooth surface. It features shallow pits with a depth of around 1 - 2 nm and the normally observable step flow pattern. The terraces of the pattern consist of steps with a height of one to two atomic layers. The density of the pits is in the range of $5 \cdot 10^8$ cm^{-2} - $1 \cdot 10^9$ cm^{-2}, which is a usual value obtained by GaN MOVPE on 2 inch sapphire templates with around 4 µm thickness. The typically observable bimodal defect size distribution is adumbrated by the different depth of pits, e.g. marked by the two arrows in fig. 4.2 a).

The surface morphology of a HVPE bulk GaN substrate overgrown by MOVPE was also investigated by AFM and is shown in 4.2 b). The overgrowth ended at the p-GaN surface with a smooth surface with step flow pattern. The offcut between the normal vector of the GaN c-plane to the normal vector of the wafer surface is vendor specified with $0.35° \pm 0.15°$ and is reflected by the aspect ratio of terrace height to lateral step distance according to fig. 3.2. Due to the absence of pit like defects the atomic steps are well aligned according to the offcut.

The bulk GaN substrate is vendor specified with a dislocation density of $< 5 \cdot 10^6$ cm^{-2} and a roughness of < 0.2 nm. After the MOVPE both values remain small, similar to the specification of the bare bulk GaN substrate. The fact that no pit can be observed on the 25 µm^2 micrograph formally results in a defect density of $< 4 \cdot 10^6$ cm^{-2}. Although this estimation demands a more thorough and statistical investigation, which was not performed here, it indicates that no further dislocations are created during the MOVPE overgrowth in a small geometric scale. In contrary, in a larger scale few pyramidal defects with a height of several 10 nm and lateral size of 100 nm - 1 µm were found. These defects are associated to larger surficial defects of the bulk GaN wafer before the overgrowth. Another source can be particles generated during growth or originating from the growth environment.

Since, a large amount of Mg is still passivated by hydrogen after the MOVPE growth, a thermal treatment has to be performed in order to remove it. In the following section the surface morphology after the hydrogen out-diffusion process will be described in more detail.

4.2 Hydrogen out-diffusion treatment

As described in Section 2.1, MOVPE grown p-GaN layers include high hydrogen content passivating the incorporated Mg-Atoms, a curse as much as a blessing though. During growth the unintenional codoping by hydrogen suppresses the p-type growth and thus the formation of compensating species. Later the hydrogen has to be removed in order to "activate" the p-type behaviour [119, 52, 120].

First observed by Nakamura et al. [121, 122] and Amano et al. [46] a hydrogen out-diffusion treatment can be applied to remove excess hydrogen from the layers. Typically thermal annealing under air or nitrogen with or without oxygen addition in a temperature range of 600 - 1000 °C is performed. Wampler et al. showed theoretical calculations for the release of hydrogen under different anneal conditions [123, 124]. The major influences for the release of hydrogen from the layers are:

- the thermal budget
- potential capping of the Mg doped layer
- and the hydrogen partial pressure in the anneal environment

It is well known and reported in literature (e.g. in [125]), that high temperature annealing degrades the uncapped GaN surface especially under low nitrogen partial pressure. The application of a capping layer (typically AlN or Si_3N_4) can suppress this degradation [126]. However, an additional overgrowth of the p-GaN surface before the high temperature anneal has to be performed for both, true-vertical and pseudo-vertical devices. Further, the hydrogen out-diffusion is suppressed by a capping layer, due to a slower diffusion inside the cap-layer.

As-grown surfaces are generally preferred for the epitaxial growth of subsequent layers to obtain defect-free interfaces. Thus a temperature below the surface decomposition temperature under pure nitrogen atmosphere was chosen, which is around 800 °C to 850 °C. The AFM image in fig. 4.3 shows the p-GaN surface of a pseudo-vertical stack after a 750 °C anneal in nitrogen atmosphere at ambient pressure for 10 minutes. The typical appearance without surface degradation, similar to the as-grown p-GaN surface shown in fig. 4.2 a), can be observed.

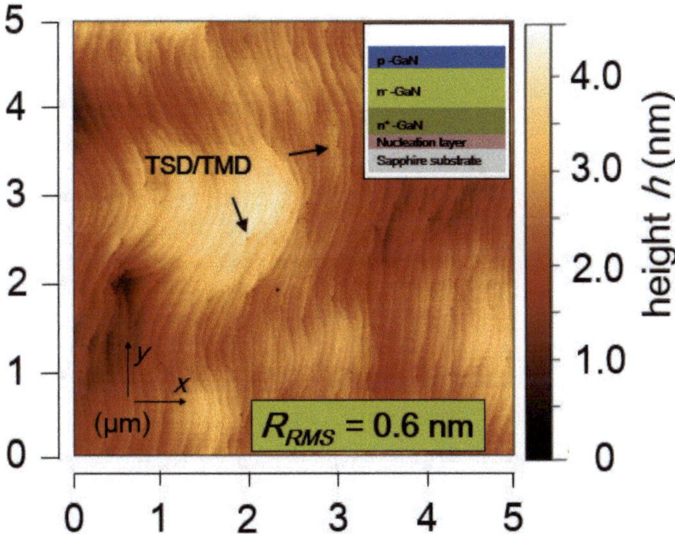

Fig. 4.3: AFM image showing the morphology of a typical sapphire based template obtained after the MOVPE growth and after a subsequent 750 °C hydrogen out-diffusion anneal.

In fig. 4.4 a p-GaN surface after a 900 °C anneal is presented. The smooth atomic terrace structure transformed into a terrace structure with a step height of several atomic layers and a higher roughness. Pit-like defects are not visible, but small droplet-like peaks are formed on the surface. Their formation is attributed to excess gallium agglomerating after the decomposition of GaN under release of nitrogen at these high temperatures. Overall the change in the surface morphology after annealing at high temperatures is not necessarily of disadvantage for the crystal quality in the following epitaxy, but degradation of the electrical properties of the upper pn^+ junction is expected.

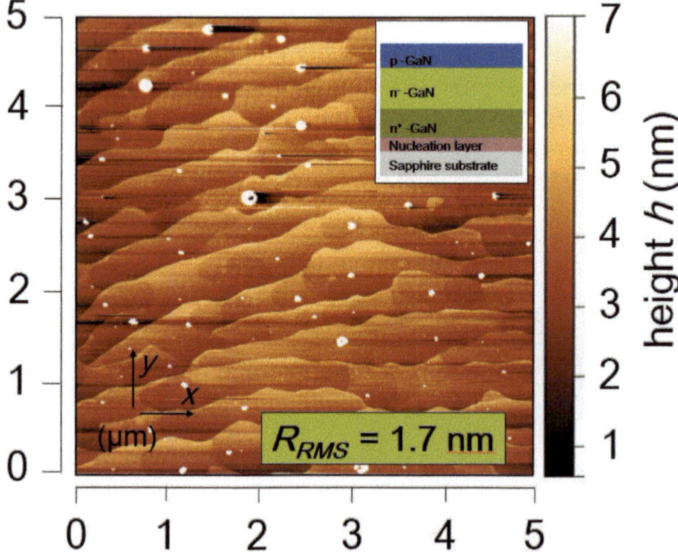

Fig. 4.4: AFM image presenting the morphology of a typical sapphire based template obtained after the MOVPE growth and after a subsequent 900 °C hydrogen out-diffusion anneal.

Another possible way to perform the hydrogen out-diffusion is annealing after the n^+-GaN overgrowth. However, in this case hydrogen has to diffuse through the overlaying n-GaN layer, but this process is reported to be insuffucient due to a suppression of H^+ diffusion in n-type GaN [127, 128]. Instead the out-diffusion of hydrogen on the sidewall of the p-GaN layer along etched areas is proposed. But due to the large lateral dimensions this approach demands much higher thermal budget for the hydrogen out-diffusion process [129]. Additionally, larger structures are not homogeneously activated.

4.3 Morphology of the n⁺-source layer grown by MBE

The entire material stack for the MOSFET requires a high quality regrowth of the n^+-source layer. After the MBE growth micrographs of the surface morphology were taken, again by AFM on the edge (fig. 4.5 a)) and in the center (fig. 4.5 b)) of the 2 inch wafer. Pits with a depth of 20 - 50 nm are found on the edge, whereas shallow hillocks with a height of around 1 - 2 nm are observed in the center of the wafer. The pitted surface causes inhomogeneities after etching and increases the effective area of the characterisation structures. Even more severe, field peaks at the pit bottom can cause inhomogeneous depletion layer distribution and early breakdown. Although the deep pits are unfavourable for the further device processing the much shallower hillocks result in a smooth surface with a roughness similar to the previous growth step.

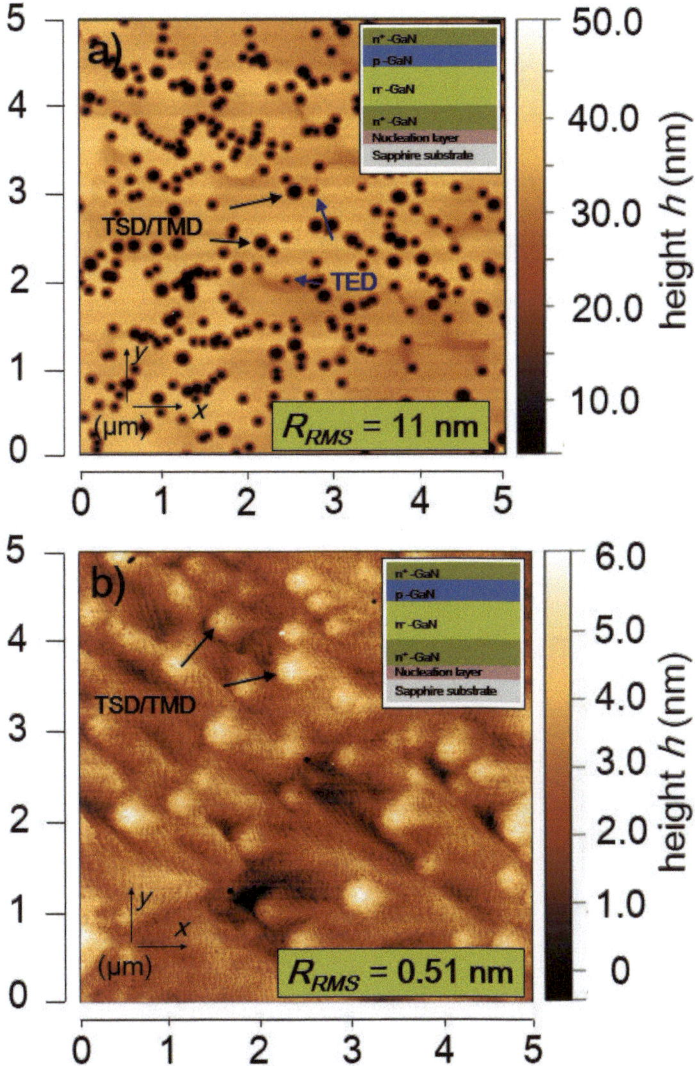

Fig. 4.5: AFM micrographs taken after MBE growth on a sapphire based GaN layer stack close to the edge of the wafer (a) and in the center (b). A bimodal distribution of deeper and shallower pits can be observed at the edge with a depth of around 50 nm and 20 nm, respectively. The step flow pattern cannot be observed under the strong height contrast of the pits, but it is present on the c-plane surface between the defects. Contrarily, hillocks with a height of a few atomic steps and a overall smooth surface are observed in the center of the wafer.

As source for the different decoration of the defects an inhomogeneity in the Ga-flux density was identified by thickness measurements over the wafers after previous growth experiments. It has to be mentioned, that the growth rate is not significantly influenced in a temperature window of about 50 K around the actual growth temperature. The growth is thus Ga-flux limited and the thickness variations permit conclusions about the Ga-flux density. In the center of the wafer the III/N ratio is higher and hillocks are obtained. At the edge in more Ga-lean conditions pits are formed [130]. Larger pits at the edge are attributed to TSD and TMD, whereas the smaller and shallower pits are attributed to TEDs. Detailed investigations on AFM micrographs reveal a screw type component of all hillocks in the center of the wafer. Hence, they are attributed to TMD and TSD, only. Additionally the pit density is higher on the edge, which further supports the finding, that the density of the large pits is roughly equal to the hillock density, defined by TSD and TMD. TED related defects are hardly observable by AFM in the wafer center, probably due to a strongly reduced decoration thereof. Nevertheless, the density of surficial defects should be similar for the edge and center, although TED are hardly observable in the later case. The entire defect density due to dislocations amounts to around $1 \cdot 10^9$ - $3 \cdot 10^9$ cm^{-2}, which is close to the value of the MOVPE grown GaN on sapphire with around $2 \cdot 10^9$ cm^{-2}. As reported in [130] the dependence of the dislocation density in the MBE layer on the used template was investigated on a bulk template with a significantly reduced defect density specified with $< 5 \cdot 10^6$ cm^{-2}. The resulting surface morphology with smooth surface and the typical step flow pattern is shown in fig. 4.6.

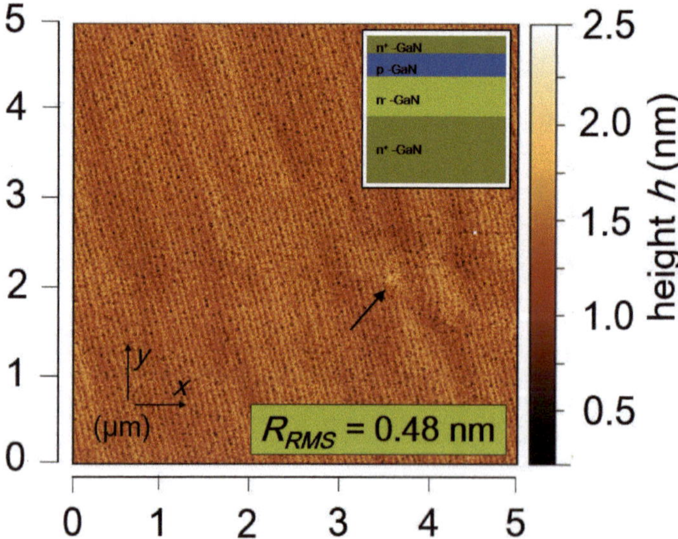

Fig. 4.6: AFM micrograph taken after the MOVPE and MBE step on a HVPE bulk GaN substrate.

Only a single feature at $x = 3.5$ µm, $y = 2$ µm can be observed, which formally correlates to a density of $4 \cdot 10^6$ cm^{-2}. Although a much thorough statistical investigation would help to examine the exact defect density it can be concluded that the defect density leading to visible surface decoration is bound to the substrate surface defect density. Hence, on low defect density templates a smooth step flow pattered surface can be obtained under different III/N ratios and temperatures. This fact verifies the high quality of the MBE overgrowth.

For pseudo-vertical layer stacks with a higher defect density a reduction or rise of the defect density was not observed. As also reported by [131, 132], the MBE decorates defects but a generation of more than $1 \cdot 10^7$ cm^{-2} cannot be observed. Addtionally it has to be mentioned, that the surface morphology is not influenced by the Si doping of the MBE layer. Similar results have been obtained for heterostructures on UID-GaN [133].

4.4 N-type doping of the functional layers

During the HVPE n-type doping can be achieved by the introduction of dichlorsilane (DCS) or by doping with a solid dopand. The later is used for Ge-incorporation. The doping of the functional layers used in this work is done by silane (SiH$_4$) for Si incorporation and bis-(cyclopentadienyl) magnesium (Cp$_2$Mg) for Mg incorporation. The growth of the MOVPE layers is discussed in detail in [134] and [135], whereas advanced studies on doping of HVPE material was done by Hofmann et al. [136]. In the following the focus is put on the MBE grown Si doped GaN used as source layer of the device. A reliable doping source inside the MBE system is mandatory to reproducibly obtain the desired Si-doping concentration.

During the MBE a Si effusion cell is used to obtain n-type growth of GaN. Characterisation of the newly installed Si sublimation source *SUSI* from *MBE Komponenten* was done first. A sample doping series was used to characterise the Si-incorporation at different Si-source currents I_{Si}. The first samples were grown at an I_{Si} of 19 A, 18 A and 17 A (not shown). Afterwards the samples were doped with an I_{Si} from 20 A - 23 A. The apparent doping concentration was evaluated by different methods, whereas for the currents below 18 A no reliable values could be obtained. The reason for that are insufficient ohmic contact formation and large deviations, due to the high debye length within the lowly doped and thin layers. The MIS- and Schottky structures were analysed in terms of $C(V)$ measurements and charge profiling described in 3.4.2. For the TLM and Hall-effect measurements $I(V)$ characterisation was used (Section: 3.4.1). The samples, which were grown first showed a higher doping concentration of around one order of magnitude. This was not expected from the extrapolation of the data from the subsequently grown layers. This effect is attributed to a start-up relaxation within the Si-source. After several 100 nm of grown Si doped material the source stabilized and the shown dependency (black guideline from 19 A to 23 A) was obtained with an incorporation of around 0.8 dec/A.

The measured electron mobility is shown in fig. 4.8, together with mobility data obtained from doped HVPE material (replotted from [136]). The obtained values from the MBE layers show similar mobilities compared to HVPE material. Additionally, literature values are plotted, which show equal or even lower values. Overall, the trend over different doping concentrations follows well the literature. Higher values in a range from $1 \cdot 10^{17}$ cm^{-3}) up to $1 \cdot 10^{19}$ cm^{-3}) speak in favour of the good crystal quality of both, HVPE and MBE grown layers.

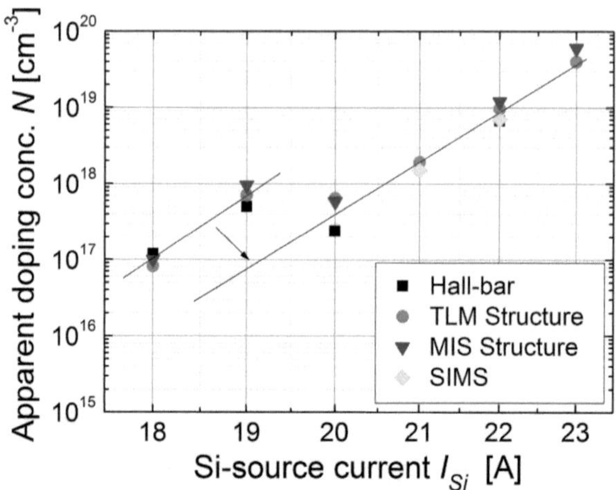

Fig. 4.7: Apparent carrier concentration in MBE GaN with Si incorporation during growth versus Si-source filament current and concentrations determined by SIMS. Estimations of the carrier concentration were made on different characterisation structures such as Hall-bar structures, MIS-structures and TLM-structures.

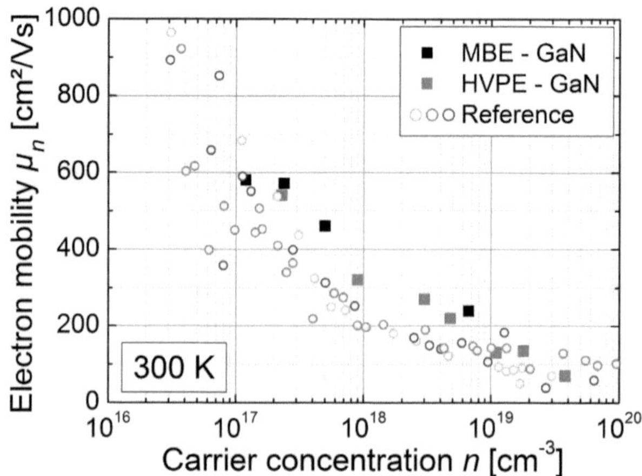

Fig. 4.8: Mobility extracted on various grown layers by MBE and HVPE (after [136]) obtained from Hall-effect measurements. Literature values are marked with open circles (pink [137], green [138], blue[139]). Additionally estimated mobility values from TLM $I(V)$ and MIS $C(V)$ measurements on the top (at $6 \cdot 10^{17}$ cm^{-3} and $2 \cdot 10^{18}$ cm^{-3}) and bottom (at $5 \cdot 10^{18}$ cm^{-3}) n$^+$-GaN layers grown by MBE and MOCVD, respectively, are presented.

The electrically active background impurity concentration in MBE-GaN of $< 1 \cdot 10^{16}$ cm^{-3} is about three times lower as for HVPE-GaN. However, above an intentional doping of around

$2 \cdot 10^{17}$ cm^{-3} these concentrations are not significant. Saturation of the mobility values is expected at doping concentrations $< 1 \cdot 10^{17}$ cm^{-3}, due to the increasing significance of the background concentration. But these lower concentrations have not been consistently characterised here.

In order to confirm the Si doping concentrations determined by electrical characterisation SIMS analyses was performed on a pseudo-vertical layer stack. The obtained data is shown in fig. 4.9. Concentrating on the deeper layers first, a concentration of $3 \cdot 10^{18}$ cm^{-3} can be observed in the n$^+$-GaN bottom layer grown by MOVPE. A concentration of $5 \cdot 10^{18}$ cm^{-3} was targeted in this layer, which is slightly higher than the concentration found by SIMS. The reduced actual doping concentration likely causes an increased drain layer resistance, which in turn is quite high for the pseudo-vertical device, anyhow. For further MOVPE growth the data will be used to readjust the dopant supply. For comparison to electrically obtained data please see Section 4.7. Since the C and O concentration is low compared to Si in this layer they are not significant in terms of doping or compensation.

Similar concentrations of Si, C and O are found in the n$^-$-GaN layer. But in contrary for the observed Si concentration of around $5 \cdot 10^{16}$ cm^{-3} the background of C and O with around $1 \cdot 10^{17}$ cm^{-3} and $3 \cdot 10^{16}$ cm^{-3}, respectively, can not be considered as insignificant. Carbon is well known as compensation agent in GaN and can partially compensate the n-type Si doping. This is further discussed in Section 4.7 with the help of fig. 4.18.

In the Mg doped p-GaN layer the Si and O concentration further drops by a factor of 2, whereas the C concentration remains almost constant. A strong compensation of the p-type GaN is thus not expected, though certain uncertainties appear from the complexity associated to the various impurities and the low thermal ionization of the Mg dopant.

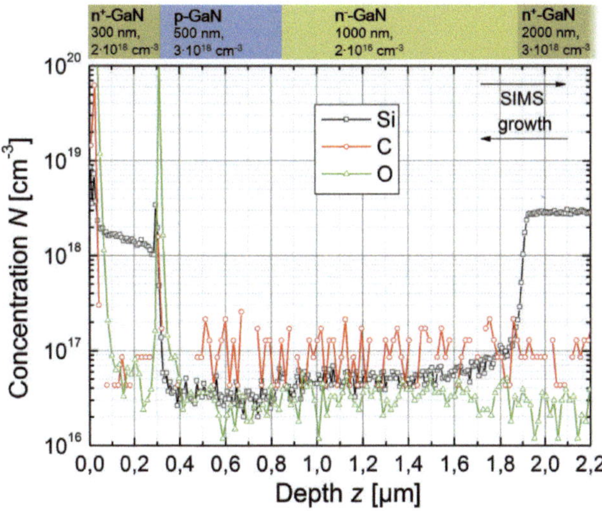

Fig. 4.9: SIMS profile of a pseudo-vertical layer stack showing the concentrations of Si, C and O from the surface up to a depth of 2.2 µm. Additionally the layer sequence is shown.

Next, the top n$^+$-GaN layer grown by MBE and its interface to the MOVPE grown p-GaN layer is focused. The sample was extensively wet chemically cleaned prior to the MBE growth. Nevertheless, a strong peak of impurities is found on the n$^+$/p-GaN interface, which indicates an increased impurity concentration on the surface of the MOVPE grown GaN before the MBE growth. For the later MOSFET operation the source-body junction is less critical, however, optimized sample cleaning and out-gassing in the UHV environment at elevated temperature should be part of further investigations to improve the interface. The incorporated doping concentration in the bulk MBE layer is close to the targeted value of $2 \cdot 10^{18}$ cm^{-3}. A shallow gradient with slightly increasing concentration towards the surface is observed. One possible reason is a thermal relaxation of the Si doping source during the growth run, but is not critical in terms of device performance. The background concentration of O in the MBE GaN layer is comparable to results shown in [93]. Similarly as for the $^+$-GaN bottom layer it can be neglected, due to the much higher and dominating Si n-type doping. The observed C concentration represents the detection limit of the SIMS setup and is not significant. In the following section the p-type doping of GaN by magnesium is discussed.

4.5 P-type GaN by magnesium doping

In fig. 4.10 the obtained doping concentration versus the dopant precursor (Cp$_2$Mg) flow during growth, determined by mercury probe capacitance versus voltage measurement is shown. In this $C(V)$-method a mercury Schottky-contact is used in order to evaluate the capacitance and doping concentration. As expected a linear dependence of the doping concentration with $2 \cdot 10^{16}$ cm^{-3} sccm^{-1} precursor flow is obtained. The used layers have been activated before these measurements.

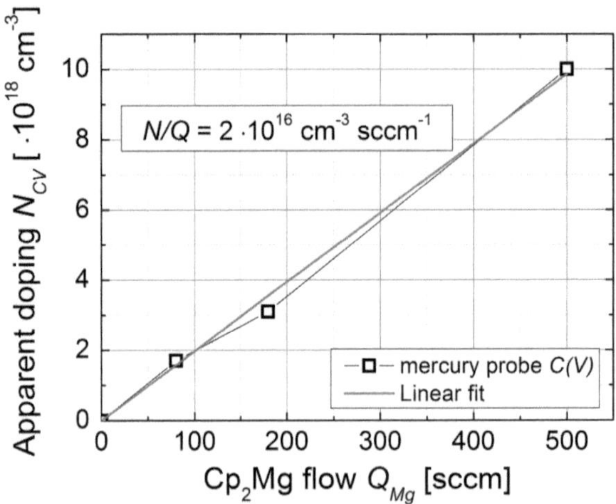

Fig. 4.10: Apparent doping concentration obtained by mercury probe $C(V)$ on activated p-GaN on a pseudo-vertical stack; a linear fit through the point (0,0) results in a doping concentration of $2 \cdot 10^{16}$ cm^{-3} per standard cubic centimetre (sccm) precursor flow

The linear trend obtained from the fit to the data indicates a reliable evaluation of the activated Mg-concentration by the mercury probe $C(V)$ method. To confirm the values obtained by electrical characterisation SIMS was performed on a non-etched surface of a sample, which was processed until the gate trench formation.

The measured Mg concentration versus depth is shown in fig. 4.11. Starting at the largest depth, the concentration in the n⁻-GaN is below $5 \cdot 10^{15}$ cm⁻³, which represents the detection limit of the Mg concentration in the SIMS system. The actual concentration can be lower, but is surely not higher in these region. At a depth of 800 nm the Mg concentration increases over the next 150 nm to around $2 \cdot 10^{18}$ cm⁻³. Thus the transition from the end of the n⁻-GaN layer into the p-GaN layer is smooth. The upper 400 nm of the p-GaN layer show a slightly increasing Mg concentration from $2 \cdot 10^{18}$ cm⁻³ to $4 \cdot 10^{18}$ cm⁻³, well compliant to the targeted concentration of $3 \cdot 10^{18}$ cm⁻³. A fast drop of the Mg-concentration as well as a very low Mg trace in the MBE GaN reveals an abrupt metallurgical n⁺p junction. Finally, on the surface of the n⁺-MBE GaN an unintentional Mg trace of around $5 \cdot 10^{16}$ cm⁻³ can be found. Since this concentration is low compared to the Si doping concentration its influence on the n-type doping is negligible. Further, surface near regions are typically influenced by surficial residuals from the sample processing. It is thus not unlikely, that the found concentration occurs from residuals and impurities generated during processing and fade out of the signal in the first 10 nm of the material. Typical sources of Mg can be chamber parts made of Al alloys, which are used e.g. in plasma reactors.

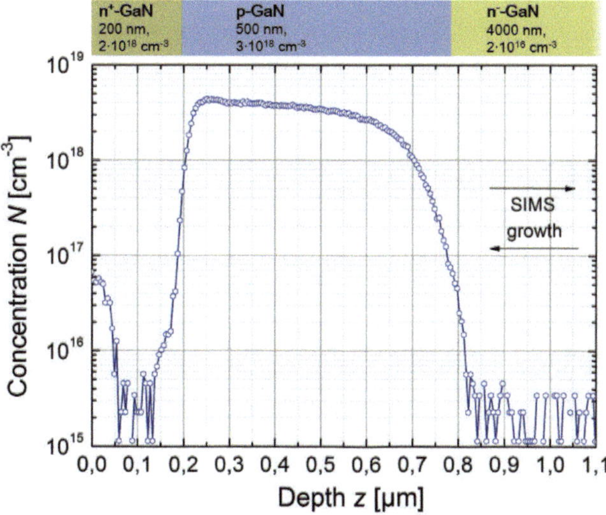

Fig. 4.11: SIMS profile of a true-vertical layer stack starting at the surface up to a depth of 1.1 µm showing the Mg concentration in the n⁺, p and n⁻ layers

The obtained concentrations by $C(V)$ charge profiling for a doping of $3 \cdot 10^{18}$ cm⁻³ match the concentration obtained by SIMS. Although only a single concentration from fig. 4.10 was confirmed, the assumption of negligible interstitial incorporation, which was made in Section 2.1 is considered as valid. Furthermore and apart from etched surfaces an extensive compensation of Mg by UID Si, O and C impurities is unlikely. The Mg concentrations obtained in this section, are thus taken as Mg_0 reference for following investigations. Thus, for the activation efficiency only thermal or field induced ionization and hydrogen passivation is significant.

4.6 Structural properties after the etching and gate module formation

Different process steps such as the gate module fabrication, the isolation and the p-GaN opening include etching with an ICP-based process. For the gate module an additional wet chemical etching step is included. In order to optimize these processes for their individual application a deeper investigation was performed and results are shown and discussed in the following.

In Fig. 4.12 the schematic cross section of a pseudo-vertical device after the gate module fabrication is shown. At this step of the process flow cross sectional images from the three different edges were taken. As can be seen, the gate trench edge has an angle of 90° with respect to c-plane of the layers. The p-GaN opening and pseudo-vertical/isolation etch result in a steepness of around 50°. All edges are covered by an ALD Al_2O_3 serving as gate dielectric and passivation. As described in chapter 3 the p-GaN opening and pseudo-vertical isolation etches have been performed by usage of a resist mask. Whereas the gate trench was etched using a hard mask based process. The initial ICP dry etch of the gate trench already results in a steeper etch flank (around 70 - 80°) compared to the resist based etch. The subsequent TMAH treatment shapes out the final 90° flank and removes plasma damaged material from the ICP dry etch simultaneously. A high quality GaN layer for the inversion channel is exposed on the surface of the trench sidewall.

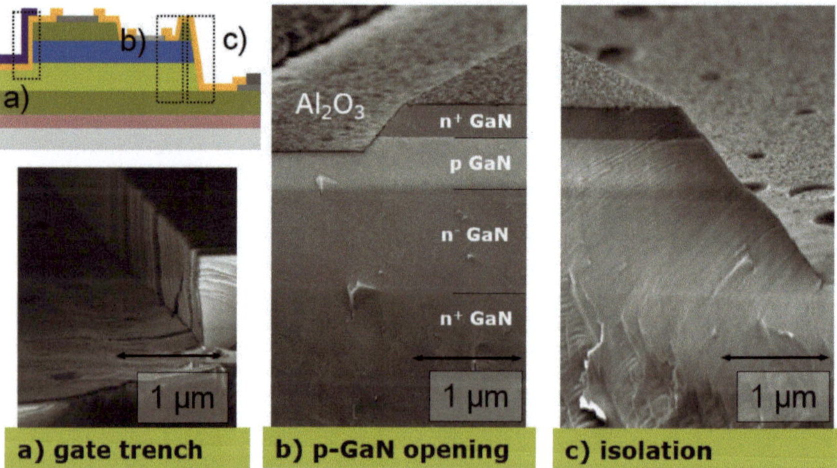

Fig. 4.12: Cross sectional SEM micrographs of the three different edges obtained after the gate module processing as indicated by the schematic device cross section. The gate trench edge is shown in a), the p-GaN opening edge is shown in b) and the isolation/pseudo-vertical etching flank is presented in c).

Two different approaches where used for the gate structuring and etching. In the first, a Si_3N_4 hard mask was deposited and afterwards structured by an SF_6 based ICP dry etch. Subsequently the GaN layer was structured by a BCl_3/Cl_2 ICP dry etch. Due to the use of a PECVD nitride as hard mask and a low selectivity of the chlorine based etching, this process was optimized for a high selectivity between GaN and Si_3N_4 with around 3:1 for low BCl_3/Cl_2 ratios. The maximum etching depth into the GaN is set by the maximum gate trench depth of around 1.5 µm considering a maximum p-GaN layer thickness of 1 µm and a 300 - 400 nm thick n-GaN source layer. Approximately 10 % overetch are considered to ensure a good channel connection to the n^--GaN drift layer. The usable hard mask thickness depends on the topological features, by previous structuring of the sample (e.g. isolation depth) and the particular layer thicknesses (e.g. n^+-GaN thickness). A hard mask thickness of 600 - 800 nm was chosen.

After the structuring of the GaN layer the hard mask is thinned and only an approximately 150 nm thick layer is remaining. This can hardly be seen in fig. 4.13 a), due to low contrast on the edge of the hard mask. The abrasion of the hard mask material and GaN during the dry etch and the selectivity of the etching recipe was evaluated by cross sectional SEM images, which were taken after the etching on test samples. After process development, two inch wafers with the desired layer stack were etched following the process sequence of the MOSFET.

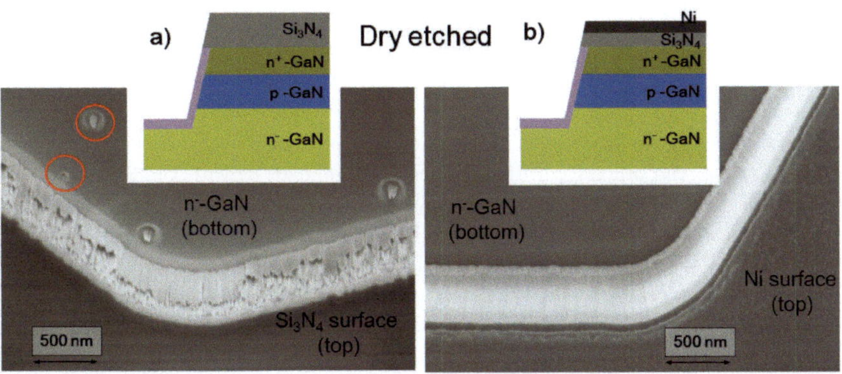

Fig. 4.13: Inclined SEM top-view and schematic cross sectional view of two trench edges after dry etching with a pure Si_3N_4 hard mask (a) and with a Si_3N_4/Ni based hard mask (b) on similar pseudo-vertical layer stacks. Plasma damage occurs at the edge flank and on the trench bottom (pink layer). Defects decorated during the etching procedure are encircled in red.

The major disadvantage using a pure Si_3N_4 hard mask is the low selectivity, which demands very thick hard mask layer. Difficulties arise due to stress formed in the PECVD deposited Si_3N_4 layer, by the hard mask structuring, back etching of the mask during the GaN etch step and finally during the removal of the hard mask after the etching steps. Thus in the second approach a combined Si_3N_4/Ni hard mask was tested. The selectivity between GaN and Ni is reported to be much higher (around 30:1) for the used chlorine chemistry [140]. This enables a thinner hard mask with less back etching and optimization of the chlorine process for surface smoothness and steep trench flanks. As can be seen in fig.

4.13 b) the surface of the trench bottom shows no features associated to defect decoration during the ICP etching. A much smoother trench edge compared to fig. 4.13 a) can be found, after the optimization of the etch recipe towards higher BCl_3 contents with the Ni hard mask.

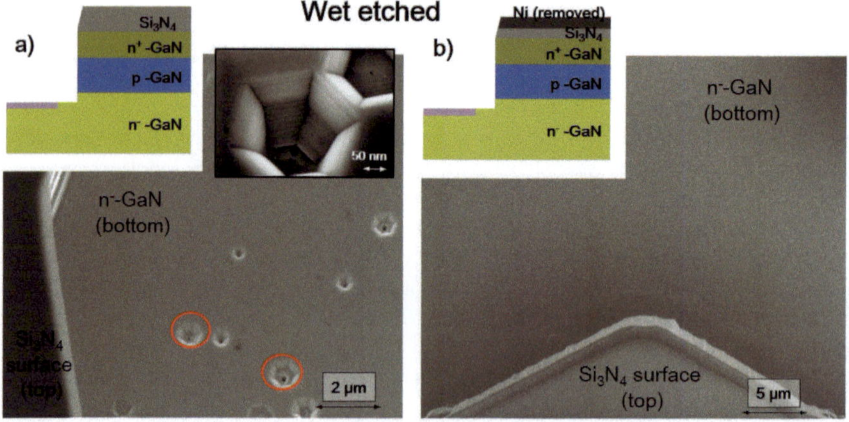

Fig. 4.14: Inclined SEM top-view and schematic cross sectional view of two trench edges after the wet etching with a pure Si_3N_4 hard mask (a) and with a Si_3N_4/Ni based hard mask (b) on a similar pseudo-vertical layer stack. Plasma damage (pink) at the edge flank is removed by the TMAH treatment. The Ni hard mask was removed prior to the wet etching.

The wet chemically etched surfaces have been studied by SEM. In fig. 4.14 both trenches are shown, in a) the process with Si_3N_4 hard mask and in b) with the Si_3N_4/Ni hard mask. Defects which were decorated already at the dry etching step (fig. 4.13 a)) got further decoration by the TMAH treatment (fig. 4.14 a)). Pits with a hexagonal shape and a depth of several 10 nm occur probably around dislocation termination lines and surficial defects. A detailed view is inset into a). In device operation these pits can lead to insufficient coverage of the gate trench by the gate dielectric. Electric field peaks under the gate dielectric in vicinity of the pits and earlier electrical breakdown of the device is promoted. The structure shown in b) shows no decoration of defects by TMAH. Thus the decoration of the defects during the wet etching is strongly dependent on the initial decoration of the defects after the dry etching step.

Next, the shape of the sidewall is discussed with the help of fig. 4.15 a) and b) for short etching times and c) and d) for long etching times. Further a comparison of m- and a-plane sidewalls is shown. Although, reports about the orientation of the trench sidewall along different planes stated a smoother sidewall for the m-plane [114, 141], the observed m-plane sidewalls showed surface stripes along the sidewall for short as well as long etching times. The a-plane surfaces, in contrary, showed a staircase, crystallite featured sidewall with small m- and c-plane facets for short etching times. For long etching times a vertically striped sidewall depending on the initial steepness is created. The difference to previously reported findings is attributed to small misalignments of the lithographic mask to the crystal directions. As consequence, on m-plane stripes are created, whereas on a-plane oriented sidewalls the obtained columnar structure, which is interpreted as alternating small surfaces with m-plane orientation and 120° rotation is formed. In the later case, the orientation offset of the lithographic mask is compensated.

For part c) and d) of the figure dark regions on the surface can be found after the TMAH treatment, which cannot easily be removed at this step of the process flow. Experiments on the sequence of the hard mask removal revealed that these residuals are created during the TMAH etch on the Ni hard mask and can be avoided by removing the Ni part of the hard mask before the TMAH treatment. This was also done before the TMAH etch shown in fig. 4.14. At this point the function of the additional Si_3N_4 beneath the Ni falls into place. It serves as protection of the previously processed structures during the removal of the Ni hard mask and the subsequent TMAH treatment.

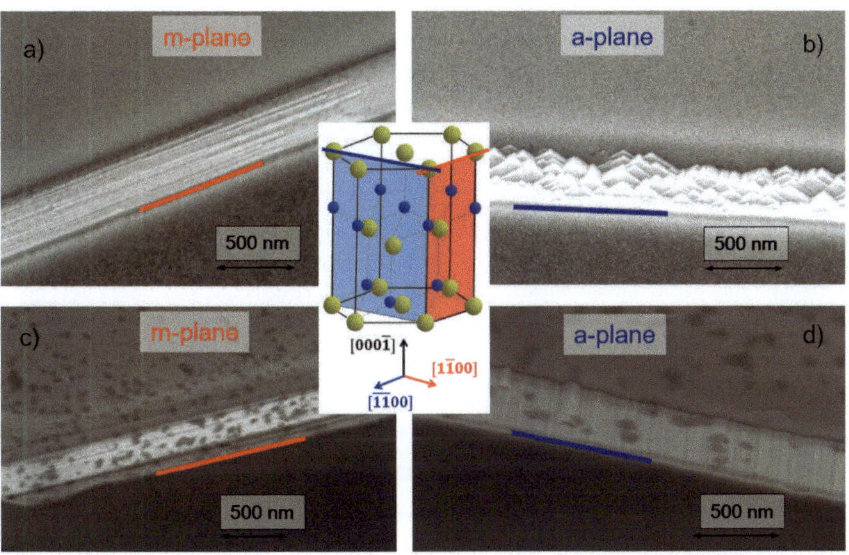

Fig. 4.15: Inclined top view of gate trench edges produced with a Ni hard mask for different wet etching times and a schematic crystal unit cell of the hexagonal GaN crystal with its respective crystal vectors. In a) and b) the etching time was around 30 min while for c) and d) it was 60 min at 80 °C. Different crystal planes are indicated by colour.

As can be seen by comparing fig. 4.15 c) and d) a small misalignment of the trench to the crystal directions on the GaN wafer a striped trench surface is obtained on the m-plane trench. A perfect alignment of the lithographic layer finally defining the trench alignment is often not possible, due to small failures during lithography and mask alignment. The etching of the a-plane sidewall is less affected by small misalignments of the etching mask, due to its different etching characteristics. Small vertical crystallite facets appear after short etching times (see fig. 4.15 b)) which are related to the etch-limiting c- and m-plane. With sufficiently long TMAH treatment the initially appearing steps formed by the c-plane of the crystal are etched until a uniformly vertical channel surface is created with vertical m-plane stripes.

After the formation of the gate trench structure and the removal of the hard mask material the gate dielectric and electrode are deposited. In order to determine the doping concentration in the n-GaN layers as well as dielectric and dielectric/GaN interface properties electrical characterisation of MIS-capacitor structures is discussed in the following section.

4.7 Electrical layer characterization

In this section results about the electrical properties of the used functional layers are presented and discussed. The data was obtained from MIS-structures processed in parallel to MOSFET structures on the same dies and wafers. Before the actual characterisation is shown, some aspects concerning the characterisation strategy are given.

Usually the doping concentration of the layers can be accessed by $C(V)$ measurements. In order to get a better inside in the electrical active doping concentration these measurements were performed on circular characterisation structures. First the capacitance versus voltage curves are evaluated in accumulation, i.e. for n-GaN in the positive bias branch and for p-GaN in the negative voltage region. The accumulation capacitance should be equal to the oxide capacitance. It can be estimated by a parallel plate capacitor model formed by the metal top-electrode, the dielectric and the accumulation layer as second electrode. The capacitance is then defined by the ϵ_r of the dielectric and its thickness. Schematic representations of the measurement structures on the respective n-GaN layers are shown in Fig. 4.16.

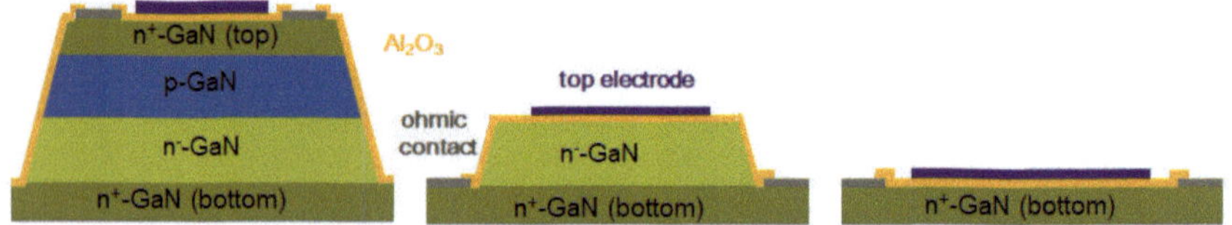

Fig. 4.16: Exemplary schematic layer stacks for the $C(V)$ measurement on different metal insulator n-GaN structures. The exitation bias and small signal voltage are applied on the top electrode, whereas the ohmic contact is kept constant at 0 V.

Typical double sweep $C(V)$ measurements started at $V = $ -6 V are shown in Fig. 4.17. Al₂O₃ with a thickness of 30 nm was used, which has a permittivity ϵ_r of approximately 9. The expected oxide capacitance per area is thus around $2.5 \cdot 10^{-7}$ F/cm² , which fits very well to the measured accumulation capacitance at around $V = $ +6 V in Fig. 4.17. The oxide capacitance is independent of the doping type, doping concentration and lateral geometry of the structure, e.g. the diameter of the top electrode.

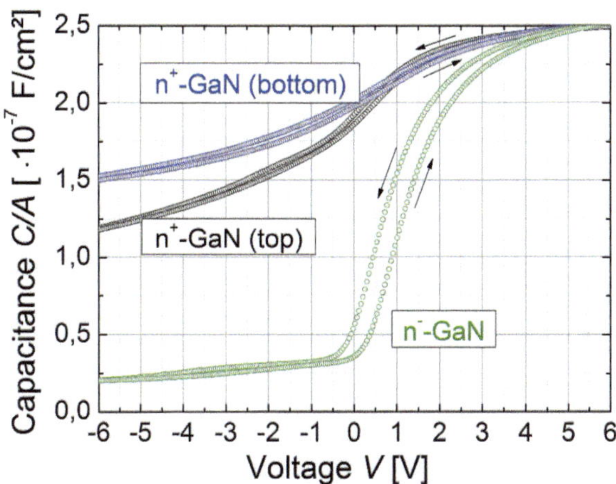

Fig. 4.17: $C(V)$ curves measured on the different n-type GaN layers using similar MIS-structures described in Section 3.4.3. The n^+-GaN (bottom) and (top) are used as drain and source layers in the MOSFET, respectively. The n^--GaN layer acts as drift layer in the MOSFET structure. The associated MIS-structure is fabricated on the level of the gate trench bottom. Capacitance values are normalized to the area and the double sweep measurement is started at -6 V.

With the approximation described in Section 3.4.2 the carrier profiles shown in Fig. 4.18 can be obtained. For the n-type material an apparent doping concentration of approximately $2{\cdot}10^{16}$ cm^{-3} for the n$^-$-GaN layer at $z > 150$ nm, $2{\cdot}10^{18}$ cm^{-3} for the top n$^+$-GaN layer and $4{\cdot}10^{18}$ cm^{-3} for the bottom n$^+$-GaN layer is obtained. These values correspond very well to the concentrations of the Si-dopants obtained by SIMS. From a comparison with Fig. 4.9 in Section 4.4 an inactive Si concentration of $3{\cdot}10^{16}$ cm^{-3} in the n$^-$-GaN can be estimated. It is likely caused by the incorporated carbon, which can act as compensation agent. The top and bottom n$^+$-GaN layers are not significantly influenced by the background C and O impurities, because their doping conc. is much higher.

The doping of the thin n^{++}-GaN capping layer on the top most layer corresponding to the source is hidden in the strongly elevated signal of accumulation typically seen close to the dielectric/GaN interface. This peak originating from the accumulation layer fades out at around $3L_D$ as described in Section 3.4.2. Further it should be noted, that the bottom n$^+$-GaN layer was accessed by a dry-chemical etching and thus the increasing signal between $z = 45$ nm and 55 nm should be taken with care. The lowest obtained concentration is considered in this case as the actual doping concentration matching the expectation from the SIMS analysis.

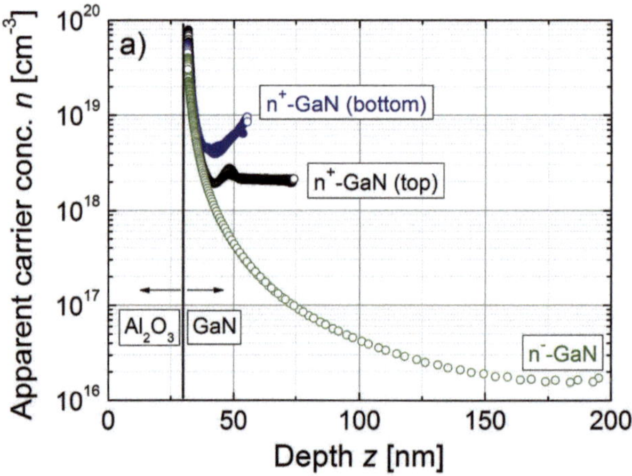

Fig. 4.18: Apparent charge profiles obtained from the $C(V)$ characteristics on the different n-GaN layers. The profiles were obtained based on the data shown in fig: 4.17. The position $z = 0$ nm correlates to the electrode/dielectric interface. The accumulation signal peak is observed at 30 nm and corresponds to the dielectric/GaN interface.

The $C(V)$ characterization on n-type GaN is usually unproblematic, due to the high activation of the shallow Si-donors. While charge profiles on n-GaN can be obtained easily the investigated p-GaN based MIS-structures showed not the expected behaviour.

A schematic representations of a measurement structure on p-GaN is shown in Fig. 4.19. The corresponding $C(V)$ measurements on differently sized MIS-structures are shown in Fig. 4.20.

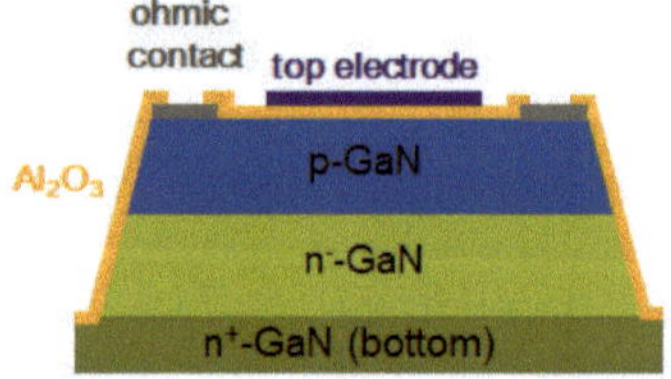

Fig. 4.19: Exemplary schematic layer stack for the $C(V)$ measurement on a metal insulator p-GaN structure. The exitation bias and small signal voltage are applied on the top electrode, whereas the ohmic contact is kept constant at 0 V.

It can be seen, that the curves are not scaling with the area of the MIS electrode. The accumulation capacitance is much lower than the expected value of $C_{ox} = 2.5 \cdot 10^{-7}$ F/cm^2 and decreases with increasing electrode area. Further, for the large diameter structures a high dissipation factor was measured (not shown), which results in a difference between the series (C_s-R_s) and parallel (C_p-R_p) modelled capacitance.

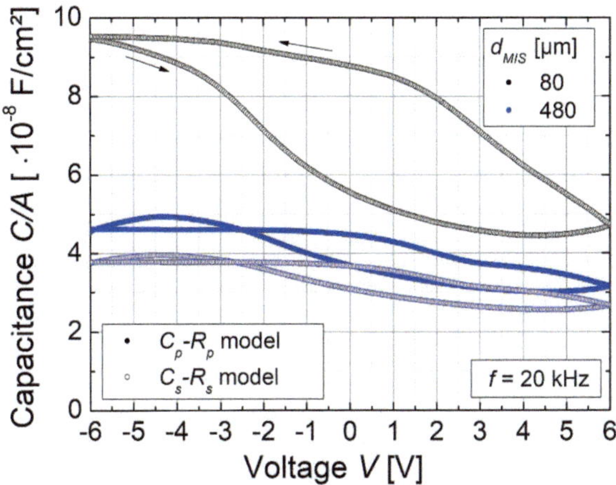

Fig. 4.20: $C(V)$ measurements on metal insulator p-GaN structures with different diameter. To curves each can be obtained depending on the applied model for a series and parallel arrangement for the resistive and reactive component. The double-sweep measurement was started at -6 V and the capacitance is normalized to the electrode area. Absent area scaling and strong hysteresis complicate the evaluation of the curves.

All together the measurements are not reliable or meaningful in the current state and a more suitable modelling of the structure has to be applied. Inductive contributions are generally considered as insignificant at the used low measurement frequencies < 5 kHz. Only a dissipative element (i.e. a resistor) within the equivalent circuit can cause a higher dissipation factor. It likely reduces the capacitance in either of the models if it is not arranged in an either pure series or parallel configuration. Further a "dead" dielectric layer formed by compensated GaN due to etching damage or a dielectric formed due to oxidation would effect the different structure sizes similarly. The effect of reduced accumulation capacitance occurs also on MIS-structures with non-overgrown, non-etched p-GaN. It is hence unclear where the reduction of the accumulation capacitance depending on the structural size originates from. An additional interface layer is ruled out as root cause for the reduction of the accumulation capacitance.

Closer investigations on the frequency dependence is shown in Fig. 4.21 for the biggest structure with 480 µm diameter in accumulation at -6 V. It can be seen, that the dissipation factor drops below 0.1 at 2 kHz measurement frequency, which correlates to a phase angle offset of < 5.7° from a pure capacitive behaviour. Again, this can not be caused by resistive contributions such as high sheet or contact resistance. Brochen et al. [57] showed that even lower frequencies can be helpful in determining the acceptor profile from $C(V)$ measurements. However, investigation on the capacitance using a *4294A Precision Impedance Analyser* from *Agilent Technologies* down to 50 Hz showed no further increase of the accumulation capacitance. Unfortunately the origin of the deviations cannot be explained so far.

Possible physical explanations could be the additional contacting of the underlying n⁻-GaN forming a pn-junction in parallel to the MIS-structure or the additional series capacitance induced by the not fully ohmic contact to the p-GaN. The potential Schottky-behaviour creates an additional space charge region under the contact, which contributes with a series-capacitance. In order to adjust the accumulation capacitance to fit with the expected dielectric capacitance a correction can be performed. A subtraction of a series capacitance or a reduction of the effective diameter can be used. Both methods would not affect the dissipation factor since their contribution would be solely non-dissipative by reduction of the over-all capacitance. In the next section a first rough correction was made by the assumption of a smaller, "effective" area, which results in an expected value for the accumulation capacitance.

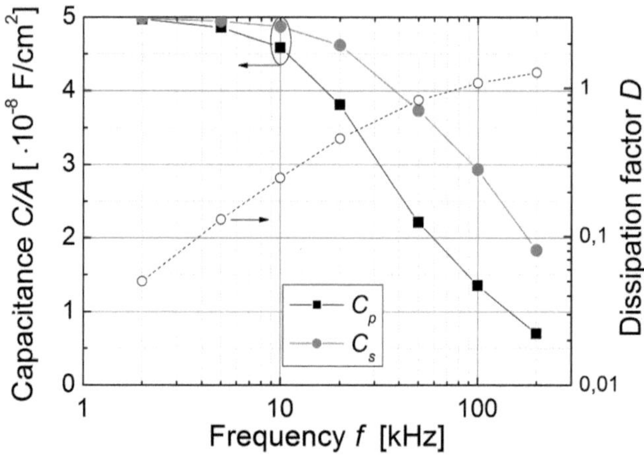

Fig. 4.21: Capacitance and dissipation factor at a voltage of -6 V over frequency for series and parallel capacitance modelling on a MIS-structure with a diameter of 480 μm

4.7.1 Gate dielectric and interface evaluation

In Fig. 4.17 a $C(V)$ curve of non-etched and dry etched n-GaN layers with a hysteresis of $\approx$ 200 mV and 500 mV are shown. For the MIS-structure on etched surfaces the back-sweep curve is shifted towards higher voltages compared to the data obtained on the MIS-structure on not-etched surface. This difference is attributed to etching damage degrading the high-k/n-GaN interface.

Similar etching was performed on p-GaN and the respective $C(V)$ curves are shown in Fig. 4.20. The hysteresis ($\geq$ 5 V) is large compared to the n-GaN structures, although the identical dielectric is used. The reasons for the large deviations have thus to be related to the biasing and carrier type in accumulation and depletion on the etched surfaces.

It is reported, that plasma damage leads to n-type behaviour of p-GaN close to the surface [142]. The formed electron enhanced layer can donate laterally electrons into the MI-p-GaN-structure under deep depletion conditions. Hence, electrons can be injected to the dielectric/GaN interface. A partial inversion layer is created in the MIS-structure. In order to investigate the dynamics of this mechanisms in more detail a sweep rate dependent $C(V)$ measurement was performed. The obtained data is shown in Fig. 4.22.

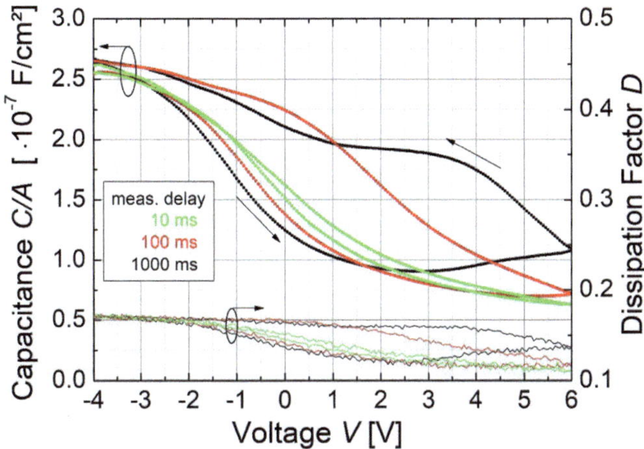

Fig. 4.22: Capacitance and dissipation factor versus voltage for a lateral MIS structure with Al_2O_3 on p-GaN with a diameter of 160 µm measured at 20 kHz starting from -4 V. The curves were corrected by a constant area factor to match the oxide capacitance in accumulation. The sweep rate was changed by adding a measurement delay time for each measurement point from 0 s to 1 s.

For slow sweep rates a strong shift of the $C(V)$ curve towards more positive voltages occurs. Additionally the onset of a partial inversion can be seen for long measurement delay times. For faster sweep rates with 0 ms and 10 ms measurement delay the $C(V)$ curves follow the trend for depletion. Reason for the time dependence can be the dynamic of the electron supply of the adjacent surface, which is only efficient at slow sweep rates. Similar behaviour occurs if the sweep only extends to lower positive bias. In this case no inversion charge can be formed under the MIS-structure and the curves behave similar as for the n-GaN structures with low hysteresis. In Fig. 4.23 a schematic band diagram of n- and p-GaN MIS-structures is presented. In case a) p-GaN trap states are initially unoccupied in accumulation and electron trapping occurs in interface traps (D_{it}) if electrons can be supplied to the structure at positive bias. A large right shift of the $C(V)$ curve results. Electrons, which were trapped at the interface can be re-emitted to the valence band dependent on the bias. This relaxation can be seen for the curves sweeped back to accumulation.

On n-GaN (b) all traps are initially filled by electrons and holes are absent. De-trapping of electrons only occurs with a very long time constant, due to the energetic depth of the interface states and the lack of empty valence band states (holes). Therefore just small shifts of the curves result.

The leakage current of the MIS-structures is not separately characterised on MIS test structures. It is discussed in conjunction with its characterisation in chapter 5 on MOSFETs and the respective gate leakage. It is shown that the OFF-state gate leakage, which represents the current through the depleted MIS-structure, is small compared to other leakage contributions. Further, the device breakdown is rather limited by the reverse operation of the pn⁻ junction.

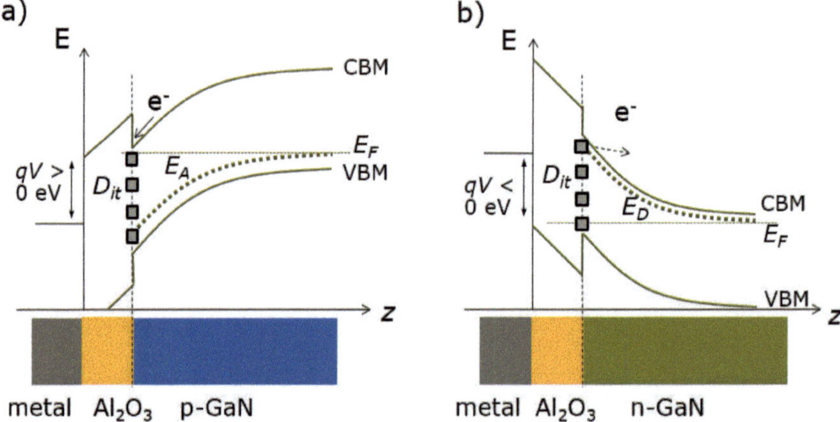

Fig. 4.23: Schematic band diagram of a MI-p-GaN (a) and a MI-n-GaN structure (b) in depletion. The direction of the electron injection into trap states located at the high-k/GaN interface is marked by the arrows.

5 Pseudo- and true vertical device operation

In this chapter different properties of the final MOSFET are characterised using different structures, which are obtained during the processing of the wafer. In order to get a better understanding of the different aspects leading to the final device behaviour under various operation modes, the usage of cooperative test structures plays an important role.

5.1 Influences of the metal-line sheet resistance

Ideally, resistive contributions from the metal layers forming source and drain contact should be negligibly small compared to the R_{DSon} of the intrinsic device. For the here processed devices with large linear width a significant influence was indicated. A closer investigation on the metal line resistance was thus performed. Fig. 5.1 presents the sheet resistances of the different metallisation layers, which were obtained on metal lines with different length.

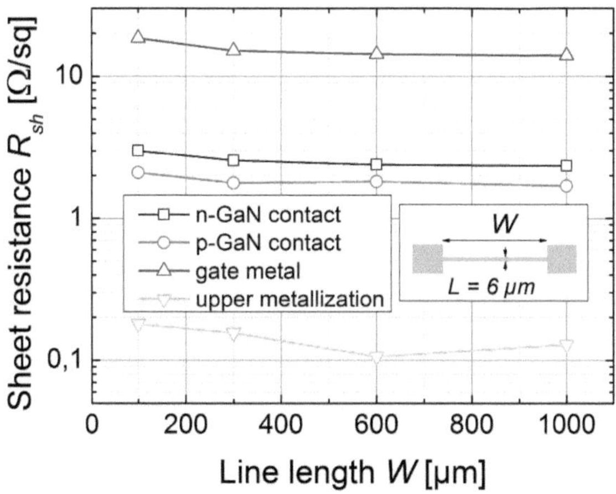

Fig. 5.1: Metal electrode sheet resistance of the different metal layers used as contacts to n-GaN and p-GaN as well as gate and higher metallization. The inset shows the used characterisation structure consisting of a metal line with two connected pads, which are not included in the calculation of the line resistance.

For small length, the obtained sheet resistances slightly increases. This is attributed to line roughness, influences of the pads and underlying layers. Values of 0.12, 2, 2.2 and 12 Ω/sq are obtained from the average of the resistances of longer lines. Elevated, metal sheet resistance, in particular for source and drain, create bias differences along the width of the intrinsic DUT. This means, that the extrinsic device resistance (R'_{DS}) at a certain position of W, changes over the width of the channel. The access of the intrinsic properties thus gets more complicated, but can be described by:

$$R'_{DS} = \sqrt{\left(\frac{R_{Msh}}{L} R_{DS}\right) W} + R_{DS}. \tag{5.1}$$

This equation can be deducted from the use of the solution for a transmission line, represented by the square root term for the characteristic resistance Z_0 and with the length of the transmission line W, which also reflects the device width [143]. The transmission line is terminated by R_{DS} in Ωmm. The corresponding circuit is shown in Fig. 5.2 b).

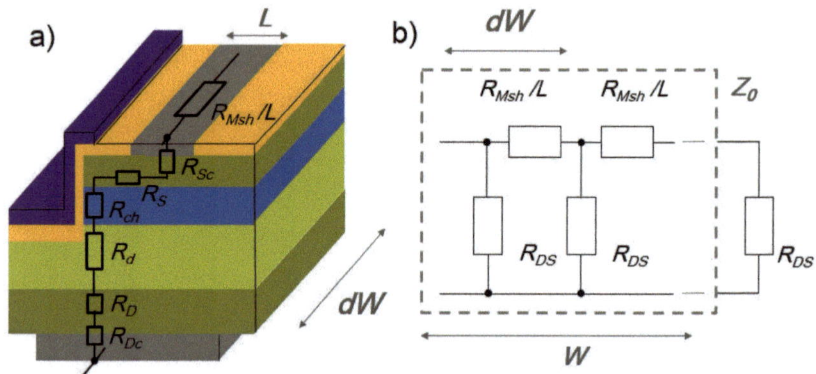

Fig. 5.2: Schematic device width-element dW (a) and equivalent circuit for the transmission line (b)

The breadth of the metal line is reflected by L. In Fig. 5.1 two examples with different R_{Msh} and L are shown. The extrinsic R'_{DS} is increased with the device width W assuming an R_{DS} of 20 Ωmm. A 10 % elevated value is already obtained at 42 µm and 290 µm width for n-contacts only and double metallised layers, respectively. It should be noted, that this assumption is only valid if R_{DS} is constant over the device width. That means, that source sided contributions of the metal line resistance shall be small.

With the re-arranged Equation (5.1) for R_{DS} the measured device resistance can be corrected to obtain the intrinsic device ON-resistance for drain sided resistance contributions. For a small source sided series resistance only a rough estimation can be achieved. With the device model explained in Section 2.4.2 the final equation set is given by:

$$V_{GS,ch} = V_{GS} - I_D \left(\sqrt{\left(\frac{R_{Msh}}{L} R_{DS}\right) W} + R_S + R_{Sc} \right) \tag{5.2}$$

for the intrinsic gate-source voltage and

$$V_{DS,ch} = V_{DS} - I_D \left(\sqrt{\left(\frac{R_{Msh}}{L} R_{DS}\right) W} + R_d + R_D + R_{Dc} + R_S + R_{Sc} \right) \tag{5.3}$$

for the intrinsic drain-source voltage. R_{DS} denotes the actual device resistance depending on the operation point without metal line resistance. Larger source-sided resistances lead to changing R_{DS} over the device width and Equation (5.1) becomes invalid.

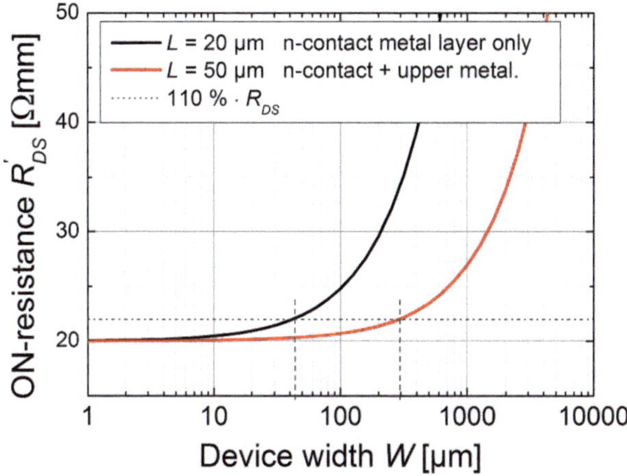

Fig. 5.3: Extrinsic device ON-resistance R'_{DS} depending on the device width with a metal sheet resistance of 2.3 Ω/sq for the n-contact metal layer and 0.12 Ω/sq for the upper metallization layer and different metal layer breadth L

Gate and body contact metal layers only have to carry leakage and displacement currents during bias change. Thus, statically their resistance is not critical. Dynamically charging currents of the gate and body junctions can get significant and additional voltage drop or an RC time constant formed by e.g. the gate capacitance and gate series resistance should be considered.

In a design suitable for power applications with a periodical hexagonal or striped pattern of numerous device cells the metal resistance contributions will be significantly reduced, due to optimized dimensions and two-dimensional current distribution. The model used here approximates a linear device structure well, e.g. as shown in fig 3.6 Section 3.3.1. The respective transfer characteristics of three devices with different width are shown in Fig. 5.4.

As can be seen, the model fits well to the transfer curves taken on similar devices with different trench width of 100, 250 and 500 µm. Since the entire sidewall of the trench forms channel area and the metal lines are surrounding the trench a width of 200, 500 and 1000 µm is used for modelling. The device is processed with upper metallization. From the previous estimation it can thus be concluded that for the 200 µm wide channel device still a ≈ 8 % increased value is obtained, which is tolerated here.

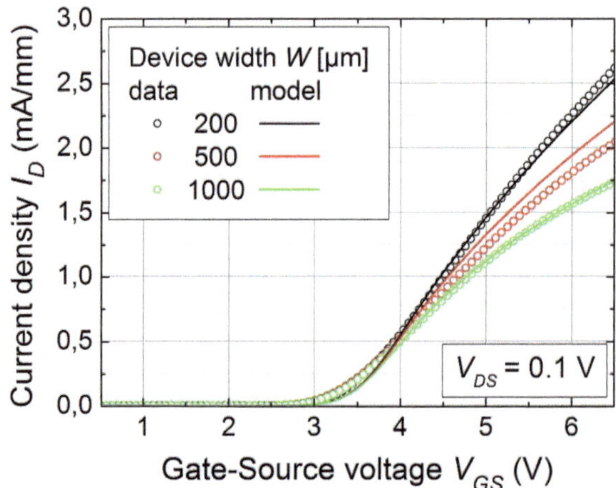

Fig. 5.4: Linear MOSFET transfer characteristics for linear devices with three different width (dotted lines). The increased series resistance contribution on the source side leads to a decreasing I_D per width at increasing V_{GS}. Modelled devices characteristics are additionally shown by full lines.

5.2 Formation and characterisation of ohmic contacts

The requirements on the used metal contact layers to both p- and n-type GaN are on the one side reasoned by their influence on the device characteristics from the electrical point of view i.e. mainly their contact resistance. On the other side boundaries are given by subsequent processing steps after the formation of the ohmic contacts and by the applied thermal annealing, which is necessary to obtain ohmic behaviour. Due to chemical surface treatments and opening of dielectric layers the contacts have to be resistant against 5 % hydrofluoric (HF) and 5 % hydrochloric (HCl) acid for at least 2 minutes. The surface roughness of the layers after the annealing is of minor importance.

In the next sections, first in part 5.2.1 ohmic contacts to the n$^+$ layers on the top and the bottom of the layer stack are discussed. Subsequently, focus is put on the contacting of the p-GaN layer in Section 5.2.2.

5.2.1 Ohmic contacts to n-type GaN

As first approach a Ti/Al/Ni/Au metal layer stack, that was developed for low contact resistance on AlGaN/GaN heterostructures [37], was annealed at 750 °C and used as potential n-GaN contact. Treatment of theses contacts in 5 % HF for 30s resulted in peeling-off of the metal layers. Subsequently, the metal layer stack had to be modified, since HF exposure occurs during the removal of the Si_3N_4 hard mask.

In a second approach a Ti/Al/Ni/Au layer stack with a thickness for the respective metal layers of 20 nm, 100 nm, 40 nm, 100 nm was used. Empirically for higher Ti contents higher contact resistances at similar contact formation anneals are expected [144, 145]. However, aging experiments on different Ti/Al based contacts indicate an improved chemical stability [146]. The application of the modified metal stack resulted in no peeling-off of contact material after the surface treatment with HF for a maximum of 2 min, which confirms this presumption. But still, a significantly longer exposure time or higher acidic concentrations can destroy the employed contact.

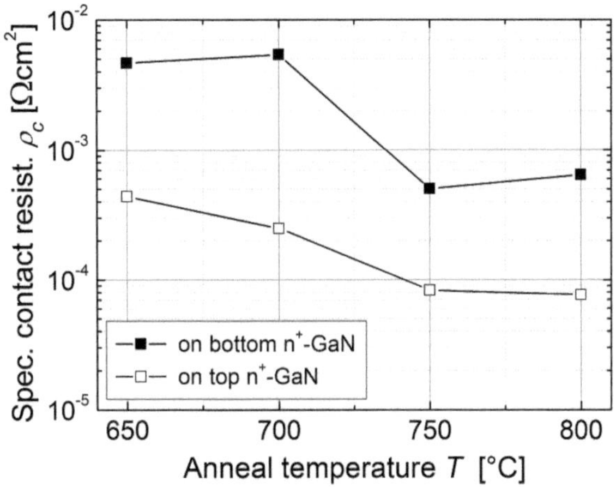

Fig. 5.5: Specific contact resistance of the Ti/Al/Ni/Au contacts after annealing under various temperatures for the source sided top n$^+$-GaN and drain sided bottom n$^+$-GaN

The specific contact resistances obtained after contact annealing with various temperatures is shown in Fig. 5.5. For both, the contacts on bottom and top n$^+$-GaN layers the specific contact resistance approaches its minimum at 750 °C. In the final MOSFET the bottom contact is related to the drain. Its contact and sheet resistance can be reduced strongly by a large area contact. For the source small area contacts are favourable in order to keep the active device area and the specific device resistance small. A higher metallisation is useful. To maintain a low thermal budget during the processing of the devices, the anneal temperature for the ohmic contacts was set to 700 °C. At this temperature the bottom n$^+$-contact resistance is still relatively high, but is not too critical for the pseudo-vertical design with usually high drain sided serial resistances. Simultaneously, an only slightly increased specific contact resistance of around 2.5·10^{-4} Ωcm^2 (related to 3.3 Ωmm) is tolerated on the top n$^+$-GaN layer. A n^{++}-GaN capping layer of 10 nm with a doping concentration of around 2 - 3·10^{19} cm^{-3} was grown on this layer. The bottom n$^+$-GaN layer obtained a doping of around 5·10^{18} cm^{-3} but was accessed by ICP dry etching, which likely causes the factor 5 to 10 higher specific contact resistance, due to plasma damage on the etched surface.

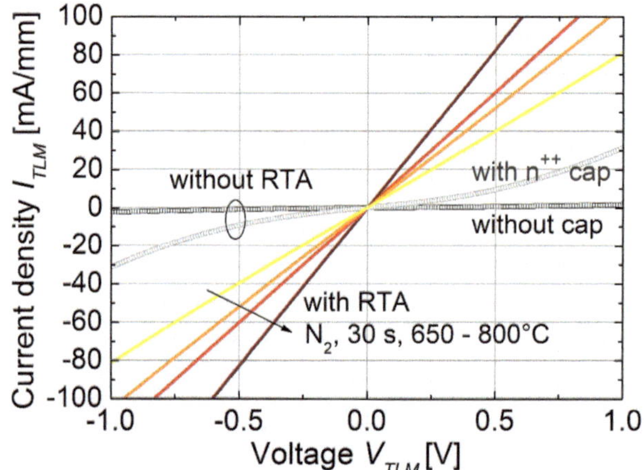

Fig. 5.6: TLM $I(V)$ characteristics of two adjacent contacts on the source layer n$^+$-GaN. The characteristics of the contact stack without (black) and with RTA at different temperatures for 30 s under nitrogen atmosphere (coloured) are shown. The distance of the contacts is 5 µm.

Figure 5.6 presents the measured current versus voltage characteristics of the different samples for the n$^+$-GaN source layer. The contact resistance for the annealed samples was successively decreased towards higher annealing temperatures from $3.5 \cdot 10^{-4}$ Ωcm^2 at 650 °C to $3.9 \cdot 10^{-5}$ Ωcm^2 at 800 °C. The linearity at 650 °C confirms ohmic behaviour, which is achieved by the application of the highly doped n^{++}-cap as upper most layer. As reference for the non-annealed characteristics the $I(V)$ curves of TLM structures with and without the highly doped n^{++}-GaN cap layer are shown in figure 5.6. A strong increase of the current for the as-deposited contacts can already be seen in case that an n^{++}-cap layer is utilized. However, the contacts are not fully ohmic.

5.2.2 Ohmic contacts to p-GaN

In this section investigations on the formation of ohmic contacts on p-GaN are discussed. Pure ohmic behaviour was not achieved neither for as-grown non-etched p-GaN layers, nor for contacts on dry etched p-GaN. However, for higher voltages on TLM structures with larger geometrical dimensions based on shadow mask processing differential ohmic scaling was measured on as-grown p-GaN material. Two different contact materials have been evaluated. The usually used Ni/Au contact stack with 20 nm of each of the evaporated metals and a sputtered TiN layer which is structured by an evaporated Ti/Pt hard mask. This hard mask can additionally be used to support the conductivity of the TiN layer, if it is not removed after the structuring.

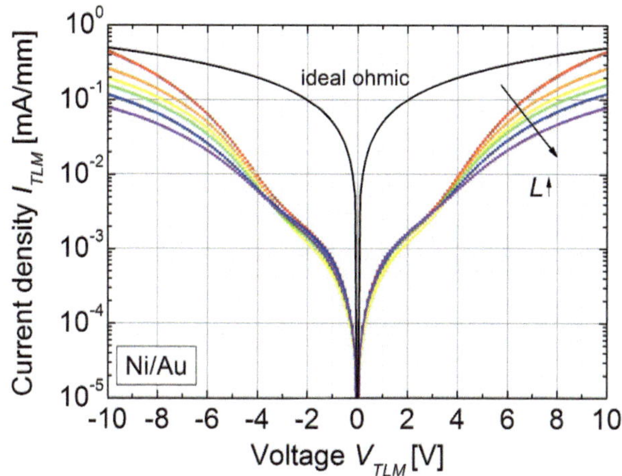

Fig. 5.7: $I(V)$ characteristics of Ni/Au TLM structures prepared by shadow mask and annealed at 500 °C in N_2/O_2. The p-GaN thickness was 300 nm with a Mg doping concentration of around $1\cdot10^{19}$ cm^{-3}.

If the formed contacts are not completely ohmic, it is not straightforward of how the resistance is evaluated. Typically for ohmic contacts the resistance is constant with the applied voltage and the absolute resistance can be used. For the here shown characteristics the differential resistance was evaluated, which allows to draw conclusions about the layer properties such as sheet resistance and active doping concentration.

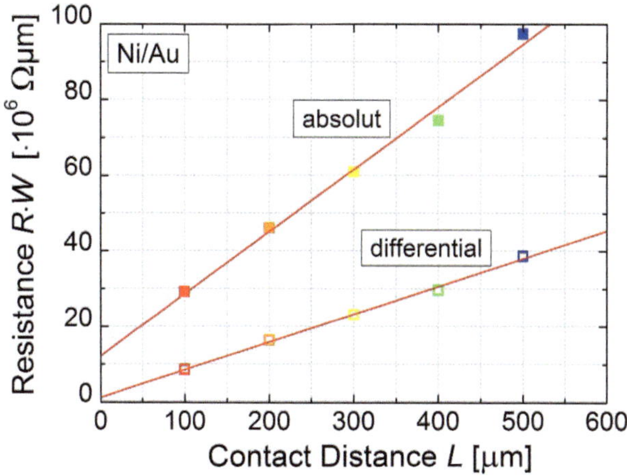

Fig. 5.8: Extracted TLM two-point resistances for different distances L of shadow mask processed Ni/Au contacts with values taken at 9 V. From the intercept and slope of the linear fit (red line) the contact and sheet resistance were extracted, respectively. Differentially extracted values are denoted by open symbols, the absolute resistance values are plotted by filled symbols.

In figure 5.7 the $I(V)$ characteristics of a TLM structure with Ni/Au metallization on p-GaN are shown. For higher voltages than $\pm\,4$ V the curves scale with the contact distance L. For the large contact distances the electric field between the contacts is still small at a V_{TLM} of around 9 V, though for non-linear curves the absolute value is hardly reliable. In figure 5.8 the resistance values from the absolute and differential resistances are compared. The $I(V)$ characteristics are shown in Fig. 5.7.

Tab. 5.1: Comparison of the obtained sheet and contact resistances for Ni/Au and TiN based contacts on p-GaN with a Magnesium doping concentration of around $1 \cdot 10^{19}$ cm^{-3}

Material	d_{p-GaN}	R_{sh}	R_c	ρ_c
	nm	kΩ/sq	Ωmm	Ωcm^2
Ni/Au	300	74	563	$4.3 \cdot 10^{-2}$
Ni/Au	500	32	695	$1.5 \cdot 10^{-1}$
Ni/Au	800	17	243	$3.5 \cdot 10^{-2}$
TiN	300	67	8100	9.7

As described in Section 3.4.3 the intercept and slope of the linear fit to the distance dependent two point resistances can be used to calculate R_c and R_{sh}, respectively. Both values are higher for the absolute resistance extraction, due to the not fully ohmic behaviour. From the differential resistances at around 9 V, for each curve the sheet and contact resistance can be estimated and both are presented in tab. 5.1.

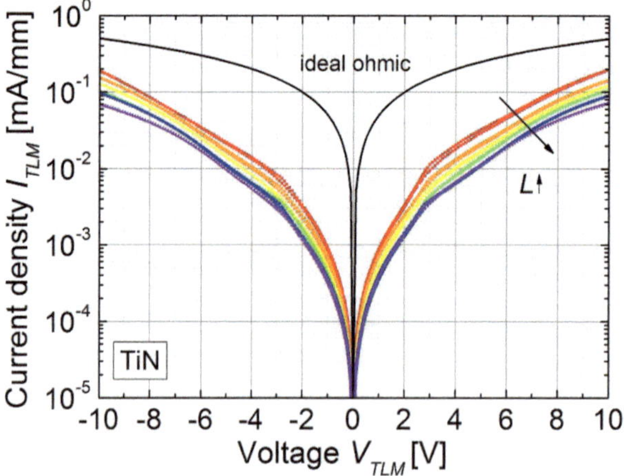

Fig. 5.9: TLM $I(V)$ characteristics of sputtered TiN contacts annealed at 700 °C in N$_2$ without substrate plasma during the sputter process. The p-GaN thickness was 300 nm with a Mg-doping concentration of around $1 \cdot 10^{19}$ cm^{-3}.

Similar to Fig. 5.7 in Fig. 5.9 the $I(V)$ characteristics of TLM structures using TiN as contact metallization on p-GaN are shown. The I_{TLM} is also scaling with L at lower bias. For comparison the differential resistances are evaluated at a similar bias point as for the Ni/Au contact. The sheet and contact resistances are summarized in Table 5.1 for both contact stacks. The obtained values for the Ni/Au contacts are well compareable to literature [126].

The specific material resistance of the GaN layer can be estimated to be around 1.5 - 2 Ωcm. Assuming a mobility of 20 cm^2/Vs [147, 134], a hole concentration of 1.5 - 2 $\cdot 10^{17}$ cm^{-3} can be estimated. Overall it can be stated, that the contact resistance of the TiN contact is about a factor of 15 - 25 higher compared to the Ni/Au based contacts. Therefore, Ni/Au contact stacks are preferred from the electrical point of view. On the other hand TiN is more stable under wet chemical treatments, during further processing. It benefits from better sticking on etched and non-etched p-GaN surfaces and provides lower metal line resistances, when a Ti/Pt mask for its structuring is used and not removed subsequently. The influence of a plasma treated p-GaN surface on the contact quality is discussed next. The p-GaN layer is accessed by an ICP-etch through the n$^+$-source layer prior to the contact formation in the process flow of the MOSFET. In Fig. 5.10 the $I(V)$ characteristics of Ni/Au TLM structures, which were deposited on plasma exposed surface are compared to contacts on untreated surfaces. The current density is reduced by more than a factor of 20 in case a plasma exposure occurred. A contact resistance is hardly extractable. Similar results can be found in case of the TiN contacts. For plasma exposure or ion etching previous to the contact deposition the contact resistance and the ohmic behaviour are strongly degraded. Therefore sputtering with substrate plasma or ICP-etching with high ion energy previous to the p-GaN contact deposition, should be avoided. For the here shown investigations large shadow mask structures were used, but similar results are expected for lithographically patterned structures under the influence of plasma etching.

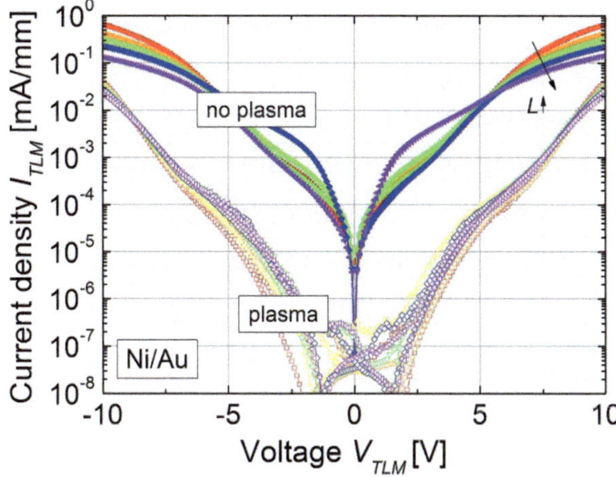

Fig. 5.10: TLM $I(V)$ characteristics of the used Ni/Au contacts without plasma influence and after substrate plasma used during TiN sputtering. The p-GaN thickness was 500 nm with a doping concentration of around $1 \cdot 10^{19}$ cm^{-3}.

In order to obtain further estimations of the mobility and the carrier concentration temperature dependent data has been measured. In Fig. 5.11 the sheet and specific contact resistance of Ni/Au contacts for different measurement temperatures are shown and the corresponding data is summarized in tab. 5.2. While R_{sh} decreases, due to a higher hole concentration at higher temperatures, R_c stays almost constant.

Tab. 5.2: Comparison of the obtained sheet and contact resistances. The ionized doping concentration for Ni/Au based contacts on p-GaN with a Mg concentration of around $1 \cdot 10^{19}$ cm^{-3} at different temperatures is estimated using a mobility of 20 cm^2/Vs and a layer thickness of 800 nm. Additionally, the estimated ionized acceptor concentration N_{A-T} by considering a temperature dependent mobility $\mu_p(T)$ is given.

T	R_{sh}	ρ_{GaN}	R_c	ρ_c	μ_p	N_{A-}	$\mu_p(T)$	N_{A-T}
°C	kΩ/sq	Ωcm	Ωcm	Ωcm^2	cm^2/Vs	cm^{-3}	cm^2/Vs	cm^{-3}
25	16.8	1.3	243	0.035	20	$2.3 \cdot 10^{17}$	25	$1.9 \cdot 10^{17}$
100	8.6	0.69	275	0.087	20	$4.5 \cdot 10^{17}$	19	$4.7 \cdot 10^{17}$
200	5.8	0.46	261	0.12	20	$6.7 \cdot 10^{17}$	13	$1.0 \cdot 10^{18}$

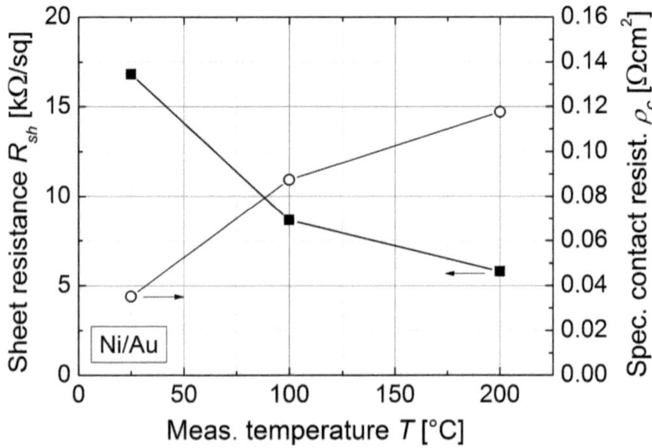

Fig. 5.11: Sheet and specific contact resistance versus TLM measurement temperature for Ni/Au contacts on a $1 \cdot 10^{19}$ cm^{-3} p-GaN layer with a thickness of 800 nm.

The carrier concentration can be calculated with an assumed hole mobility. In Fig. 5.12 the calculated hole concentration is plotted against the temperature assuming a temperature independent mobility of $\mu_p = 20$ cm^2/Vs [134]. As reference the modelled ionized acceptor and hole concentration with three different ionization energies and a residual N_D of $2 \cdot 10^{16}$ cm^{-3} were calculated, as described in Section 2.1. The data agrees well with the model for 300 K and the deviation for the data points at higher and lower temperature can be explained by the assumption of a constant mobility. If a temperature dependent mobility $\mu_p(T)$ is considered, the data matches well the model with an approximate acceptor activation energy of around 195 meV and a Fermi level of 170 meV above the valence band. The mobility varies by a factor of around 2 in this case, which is reasonable for the used temperature range. Still some uncertainties remain, since the absolute mobility values were not measured.

Additionally the conductivity of the p-GaN layer can be estimated for different temperatures. Due to the compensating nature of the deep acceptors for the shallower background donors at low temperature below around 130 K the layer gets semi-insulating (SI). The ionization of the acceptors is not linked to thermal ionization but to the donor concentration in this case. No carriers are available, which could contribute to conductivity. Above 130 K additional acceptors are thermally ionized, which lead to the generation of holes, and thus to a p-type behaviour of the layer. Indications for this behaviour was found by $C(V)$-measurement on MIS-structures measured at 100 K, 200 K, and 300K. At 300 K usual $C(V)$ curves were observed. With a decrease of the temperature to 200 K a decrease of the capacitance is found. At 100 K a very low capacitance signal close to the noise level is observed. The reason for this is likely the loss of the ohmic contact and the absence of carriers contributing to the capacitance.

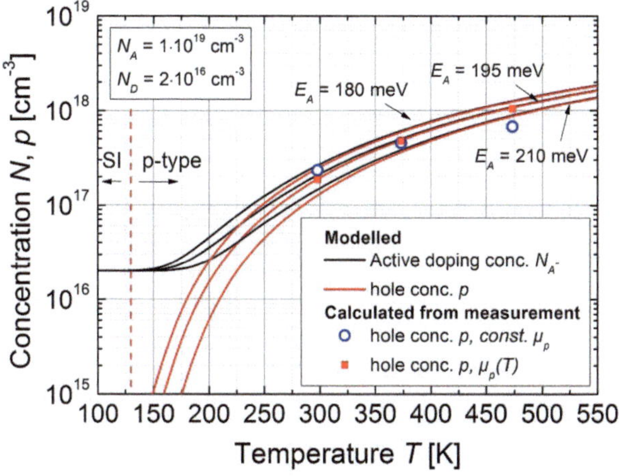

Fig. 5.12: Ionized (active) doping concentration and hole concentration from theoretical modelling with an acceptor concentration of $1 \cdot 10^{19}$ cm⁻³ and a residual donor concentration of $2 \cdot 10^{16}$ cm⁻³ versus temperature according to tab. 5.2. Different activation energies from 180 meV to 210 meV for the acceptors and 25 meV for the donors were used.

In literature a discrepancy is reported for the activation energy from electrical characterisation (150 - 180 meV) [57, 148] and photoluminescence data (190 - 240 meV)[56, 149]. A reduction of the activation energy dependent on the ionized dopant concentration is proposed as depicted by:

$$E_A = E_{A0} - \alpha N_{A^-}^{\frac{1}{3}} = E_{A0} - \alpha(N_D + p)^{\frac{1}{3}}. \tag{5.4}$$

Refering to [148] the factor $\alpha = 4 \cdot 10^{-5}$ meVcm was experimentally estimated. Due to residual donors leading to increased ionization, N_{A-} can be substituted by $N_D + p$. However, for the layers used here the unintentional n-doping is assumed to be in the lower 10^{16} cm^{-3} range, which is low compared to p at 300 K. Assuming an E_{A0} of 220 meV and $N_{A-} = 5 \cdot 10^{17}$ cm^{-3} at 300 K an ionization energy of around 195 meV fits the presented data best. Further discussion can be found in [150, 151].

5.3 The pn⁻ body diode

In comparison to the true-vertical diode characteristics, which are shown in Fig. 5.14 in the next section the characteristics of similar processed diode characterisation structures on a pseudo-vertical layer stack are discussed. The respective $I(V)$ curves are shown in Fig. 5.13. The behaviour of these diodes deviates drastically from the true-vertical ones. For pseudo-vertical diodes, the OFF-current is 2 - 3 decades higher and the ON-current is reduced by 3 - 4 decades. The ON/OFF ratio is about 10^5. Moreover, the current at higher V_{pn} does not scale with the area, which indicates non-area dependent resistance contributions probably stemming from the pseudo-vertical connection of the cathode sided n-GaN contact. Threading dislocations acting as current path in OFF- as well as ON-state should also be considered as sources for deviations. The ideality factor of the pseudo-vertical diode is in the range of 3 - 5.5 for the differently sized circular test structures. This relates to a slope of 180 - 330 mV/dec, which further points towards a series resistance contribution to the pn-junction.

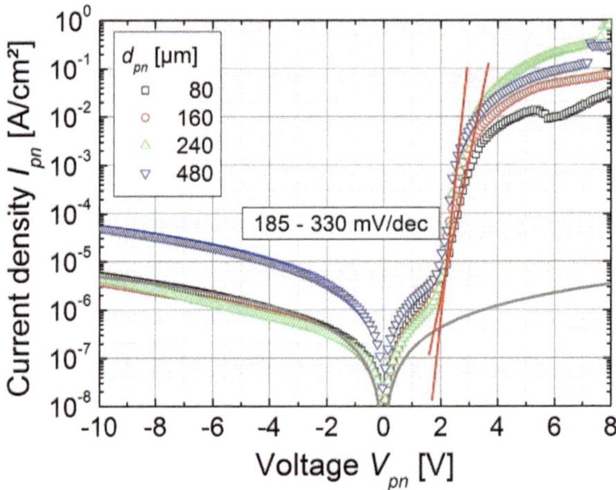

Fig. 5.13: Forward $I(V)$ characteristics of extrinsic pseudo-vertical pn⁻-diode test structures with different diameter from 80 to 480 µm. A ideality factor of 3 - 5.5 can be estimated. The gray curve shows a power law dependent current fit with an exponent of 1.5.

Overall, high performance of pseudo-vertical pn-diodes could neither be obtained in the MOSFET (not shown) nor in separated diode test structures in ON-state. In the low voltage OFF-state the curves are in most cases similar in shape and leakage current density. But, for a few devices sporadically higher OFF-leakages can be observed, e.g. for the diode test structure with $d_{pn} = 480$ µm in Fig. 5.13. Generally the leakage current depends on small negative bias in a power law with an exponent in the range of 1.5 to 2. A calculated example curve with an exponent of 1.5 is shown in gray in Fig. 5.13. This hints to surface or space charge limited current contributions as leakage sources, which were introduced in Section 2.3.1. A closer discussion in comparison with a pseudo-vertical MOSFET in OFF-state is given in Section 5.4.3.

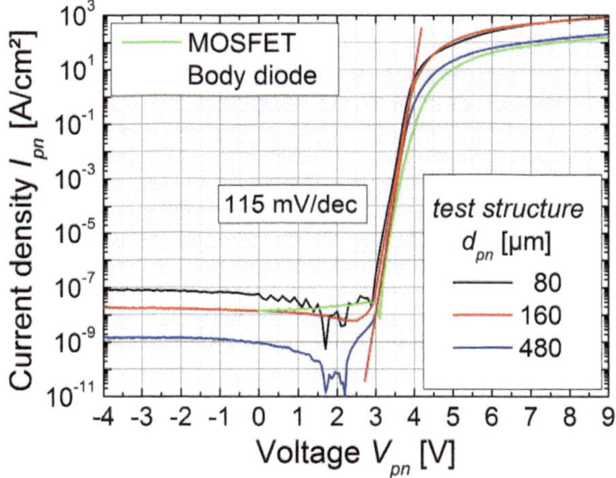

Fig. 5.14: Forward $I(V)$ characteristics of true-vertical pn⁻-diode test structures compared to the intrinsic true-vertical MOSFET body diode; reduced area related ON-current for the larger MOSFET and Diode with 480 µm diameter is observed.

Next, true-vertical diodes are discussed. The forward characteristics of the body diode on a true-vertical stack for differently sized pn diode characterisation structures is compared with the intrinsic body diode of the MOSFET in Fig. 5.14. While similar ideality factor of the diodes with n = 1.9 can be extracted, a variation of the ON-state current density is observed.

For the circular structure with a diameter of $d_{pn} = 480$ µm and for the large MOSFET body diode a reduced current is observed. This is attributed to series resistance contributions from the measurement setup and from the layer structure of the larger test structures. For the smaller diode structures an ON-current of 1 kA/cm² and ON/OFF ratio of 10^{11} was observed. Further a differential ON-resistance of 5 mΩcm² for the smaller structures and 20 mΩcm² for the large as well as for the MOSFET pn⁻ diode was extracted. The OFF-state current close to $V_{pn} = 0$ V is not scaling with the device area. This indicates that the OFF-state current is rather limited by surface leakage currents than by the area dependent junction reverse current. From the minimum current level at around $V_{pn} =$

2 V it cannot be excluded that additional carrier generation takes place, although the structures were measured in the dark. Energetic deep states in the band gap can form recombination/generation centres within the space charge region acting as generation centres. Additionally this can explain the exponential slope of the curve of approximately 120 mV/dec, which hints towards a SRH recombination mechanism. In Fig. 5.15 the smallest diode test structure with a diameter of 80 µm is shown at different forward current densities. Light is emitted around the inner opaque metal contact with changing color and brightness. From the band gap of GaN and the Fermi level of several 10 meV below the conduction and above the valence band for n-type and p-type GaN, respectively, a emittance wavelength in the range of 365 - 415 nm (3.0 eV - 3.4 eV) is expected, which is violet up to the UV-A band. The emission of green and white light thus indicates a more broad spectrum of emitted wavelength.

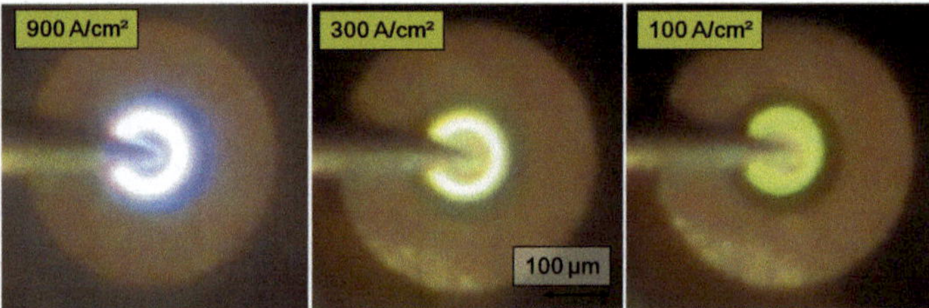

Fig. 5.15: Microscopic dark field (DF) image of the diode test structure with d_{pn} of 80 µm at different current densities; the outer ring of the diode is not connected since it is used on pseudo-vertical stacks only.

One reason for the broad emission spectrum can be related to energetically deep states inside the material as well as dislocations or luminescence in the deeper GaN layers, at interfaces as well as at the surface. Typically, heterojunction based LEDs with multiple quantum-well structures are used in lighting or signaling applications [152]. Although, from an application point of view the light emission from these elementary pn-junctions is not of highest interest and the emission is not deeper characterized from a spectroscopic point of view, it nicely illustrates the functionality and tunability of the diode.

In the following the OFF-state of the true-vertical pn⁻ junction is deeper investigated. First an $I(V)$ characterisation of differently sized diode test structures was performed by a single sweep from 0 V to high reverse voltages. The hard breakdown voltage was determined to be in the range from 360 - 400 V, which results in a peak electric field at the metallurgic junction of ≥ 2.1 MV/cm. The drift layer and body layer doping concentrations are about $3 \cdot 10^{16}$ cm⁻³ and $1 \cdot 10^{19}$ cm⁻³, respectively. The drift layer thickness is 4 µm. The depletion region is still in a triangular shape with a depth of ≈ 3.5 µm not reaching the highly n-doped drain layer under these conditions.

At low negative bias a very low leakage current close to the resolution limit of the setup is measured. Below -30 V a stronger increase in the leakage current limits the reverse operation of the diode. For a voltage below approximately -130 V the current follows an power law dependency with an exponent of 4. Similar to the pseudo-vertical junction this can be reasoned by a space charge limited or surface related leakage current. For higher negative voltages the further increase of the leakage current shows an exponential behaviour, which is more likely to be a trap assisted mechanism. Both dependencies are fitted to the curve of the diode with a diameter of 480 µm in Fig. 5.14. Investigations on different geometries and temperature dependent measurements could help to further analyse the source of the exponential leakage contribution.

Additional investigations were made by double sweep measurements at increasing maximum voltages. Compared to a single sweep breakdown measurement the sequential measurement to successively higher voltages leads to breakdown at lower reverse voltage indicating degradation of the structure during the high voltage application. On the other hand the resolution limited current density below $1 \cdot 10^{-7}$ Acm^{-2} is expanded up to -80 V, which corresponds to a stronger suppression of the leakage for low reverse voltages and after previous high voltage exposure.

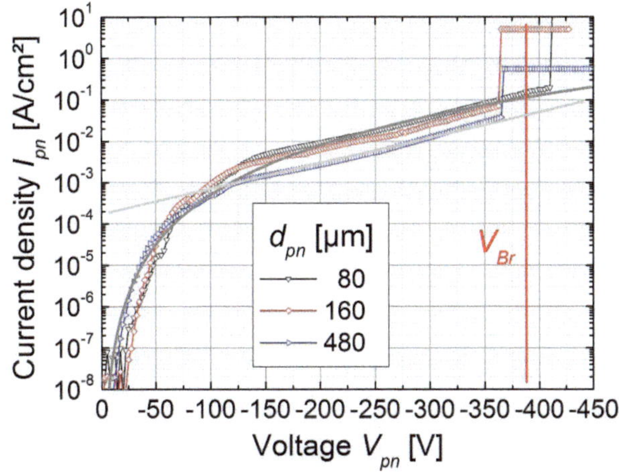

Fig. 5.16: Single sweep breakdown measurements of true-vertical pn⁻-diode test structures with different size showing a breakdown voltage of around -380 V. The dark gray line represent a fit to the current below -130 V with a power law dependence ($I \sim V^4$). The light gray line represents an exponential current voltage characteristic for higher negative bias ($ln(I) \sim V$).

In summary, potential influences resulting in a reduced breakdown voltage can be locally increased electric fields at the edges of the isolation of the structures by trapping effects or dislocations, which still should occur with a number of ≈ 250 for the diode with a diameter of 80 µm, assuming a dislocation density of around $5 \cdot 10^6$ cm^{-2}. Additionally, surface related leakage path have to be considered, as discussed before. Their influence is further discussed in Section 5.4.3. The application of a field plate (data not shown) results not in major changes of the characteristics, which hints even more into the direction of the passivation as potential weak point.

5.4 MOSFET operation

In the matter of clarity this section is divided in two parts. The first part handles with the
ON-state and turn-ON operation, such as the threshold and subthreshold region as well as
output, and transfer characterisation of the device. Subsequently, the OFF-state in terms
of device breakdown and leakage is discussed.

5.4.1 ON-state and turn-ON operation

At the beginning of this section the output (Fig. 5.18) and transfer characteristics (Fig.
5.17 and 5.23) of a pseudo-vertical device are discussed. The transfer characteristics
$(I_D(V_{GS}))$ are shown in Fig. 5.17 with linear I_D scale. As described in Section 2.4.3
from the transconductance g_m the mobility can be extracted. It is shown in the same
graph having maximum values of around 38 cm^2/Vs.

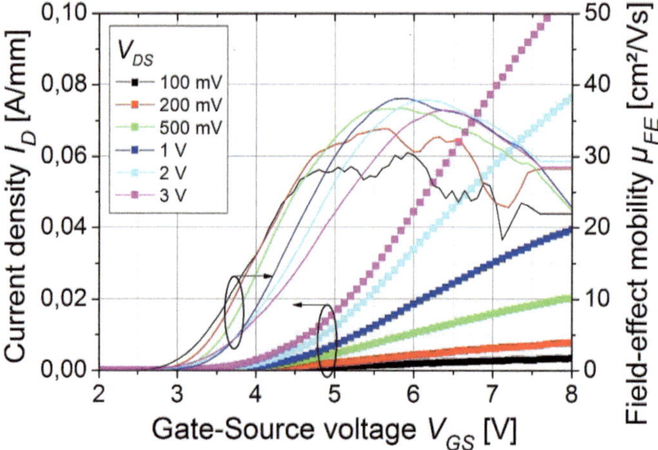

Fig. 5.17: Linear transfer characteristics $(I_D(V_{GS}))$ and field-effect mobility from $V_{GS} = 2$ V
to 8 V of a pseudo-vertical MOSFET at various V_{DS}. The current density is shown
with respect to the gate or trench width. A maximum current density of 110 mA/mm
is reached for $V_{DS} = 3$ V.

The extracted effective mobility from the output characteristics of the same device is presented in Fig. 5.19. A maximum value of 67 cm^2/Vs can be extracted, which is significantly
higher than the maximum value obtained from the transconductance extraction. The reason for the difference is related to the operation point. For the transconductance method
the maximum value is extracted at rather low carrier concentrations in the channel, where
the potential along the channel cannot be considered as constant.

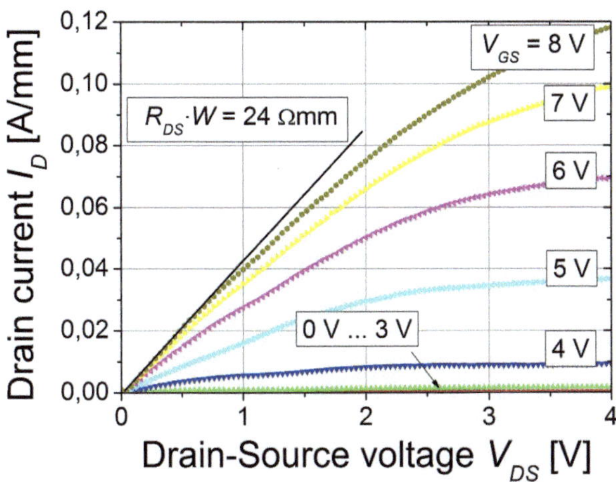

Fig. 5.18: Output characteristics $(I_D(V_{DS}))$ of a pseudo-vertical MOSFET at various V_{GS} with one volt step size. An ON-resistance of 24 Ωmm can be derived on the curve for highest V_{GS} and at low V_{DS}.

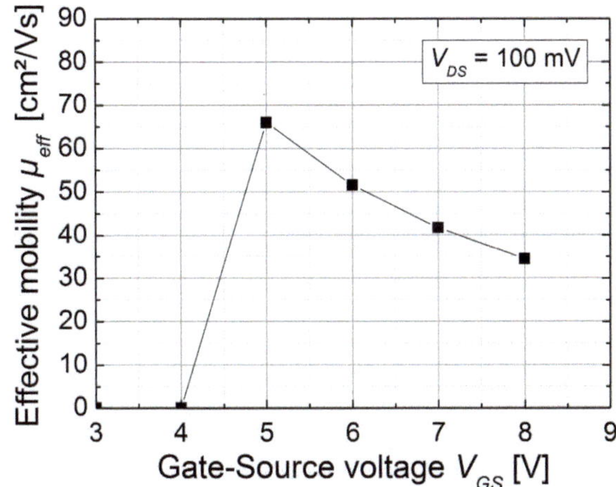

Fig. 5.19: Effective mobility versus gate-source voltage extracted from the output characteristics at V_{DS} = 100 mV for different V_{GS}.

In general the used approximations do not include series resistance contributions from the source and drain contacts and layers as well as the drift layer of the device structure. Therefore, the extracted mobility values represent lower boundaries. In Table 5.3 the values for different series resistances are summarized. Due to the small device width of 200 µm the metal line resistance is neglected.

Tab. 5.3: Series resistance contributions for the presented device (fig 5.20); R_{Sc}, R_D and R_{Dc} were estimated from TLM measurements. R_d is calculated from the specific material resistance in combination with the dimensions of the gate foot (trench bottom) surface and R_{ch} is finally calculated by subtraction of all other contributions from the R_{DS}, which is obtained from the output characteristics in Fig. 5.18. All values are given in Ωmm.

R_{Sc}	R_S	R_d	R_D	R_{Dc}	R_{ch}
4.5	2.2	1.2	7.1	1.8	8.2

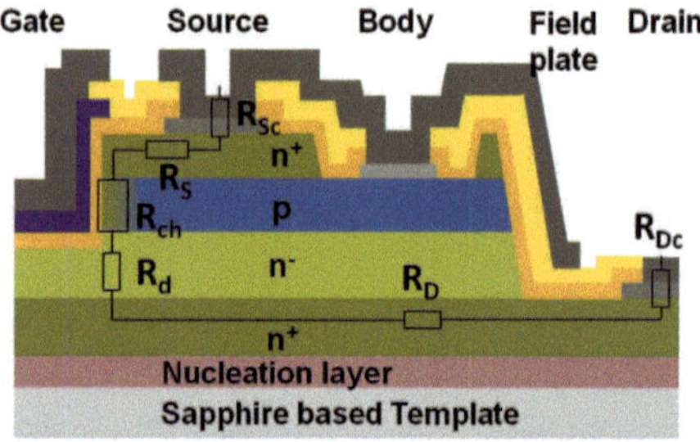

Fig. 5.20: Schematic cross section of a pseudo-vertical MOSFET with its series resistance contributions.

Using the previouly described model for the MOSFET according to Section 2.4.2 and Section 2.4.3 a channel mobility estimation can be made. Series resistance contributions according to Table 5.3, a body capacitance factor of $b \approx 3.3$, $C_{ox} = 2.5 \cdot 10^{-7}$ F/cm^2 and a gate width and length of 200 µm and 0.8 µm are used, respectively. A mobility of around 100 cm^2/Vs is obtained, which is slightly below previously reported estimations on devices with series resistance contributions [153, 154].

In order to compare the mobility values extracted on a vertical channel to values obtained from a lateral geometry a device as shown in Fig. 5.21 was investigated. The gate length is 12 µm forming a long n-channel device with a total ON-resistance dominated by the channel. The device width is 100 µm. Transfer characteristics of the device are shown in Fig. 5.22 for different drain-source voltages. The calculated field-effect mobility is about 75 cm^2/Vs. The threshold voltage is reduced by around 1 V compared to the vertical arranged MOSFET, which indicated additional positive charges on the high-k/GaN interface. Likely their formation is linked to n-type doping induced due to the plasma damage during ICP-etching onto the p-GaN, which is performed in order to access the p-GaN layer after the MBE overgrowth. Due to the plasma etching perpendicular to the p-GaN surface a much stronger degradation, especially of the channel mobility was expected.

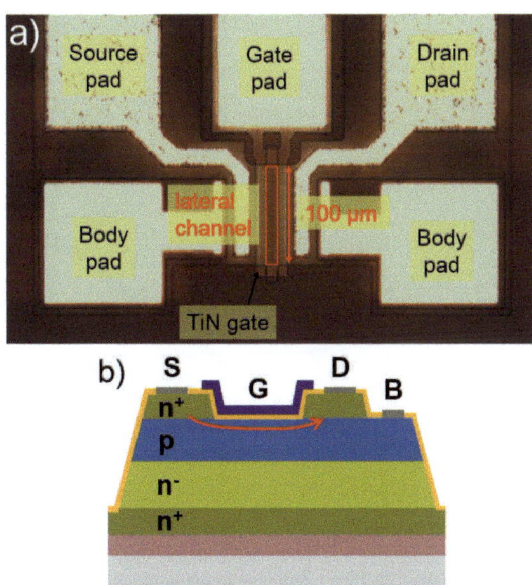

Fig. 5.21: Bright field optical microscopy top view (a) and schematic cross section with a single body contact (b) of a lateral n-channel MOSFET with 12 µm gate length and 100 µm width

With the additional information from the lateral MOSFET the mobility values of the vertical one seem relatively low, since the vertically formed channel is additionally wet chemically treated in order to remove plasma damage from the dry etch. The reason for the apparent discrepancy can be caused by much higher series resistances in the vertical MOSFET. Additionally, the better defined geometry in the lateral design such as the long channel can result in higher mobility values. The previously discussed shift of the threshold voltage, further indicates the creation of slow, positive trap charges in surface near regions. Turning back to the MOSFET with vertically aligned channel, methods for the extraction of trapped charges effectively located at the high-k/GaN interface are discussed in the following. In Fig. 5.23 the transfer characteristics are shown using a logarithmic current axis. The gate-leakage current is shown with equal scaling. It is limited by the resolution limit of the measurement setup for all bias conditions and is much lower compared to the OFF-state drain current. From the linear fit to the subthreshold region a subthreshold swing of 200 mV/dec can be extracted at low V_{DS} in the linear operation regime. The value represents a body capacitance factor of around 3.3 according to (2.37) in Section 2.4.4. With $C_{ox} = 2.5 \cdot 10^{-7}$ F/cm^2 and by neglecting interface trapping a depletion capacitance of $5.8 \cdot 10^{-7}$ F/cm^2 is extracted, which allows the estimation of the doping concentration to be around $1.7 \cdot 10^{19}$ cm^{-3}. In this device the used target Mg incorporation during growth of the p-GaN was $3 \cdot 10^{18}$ cm^{-3}, which is significantly lower. Hence, very likely the presumption of negligible interface trap charges is not valid here.

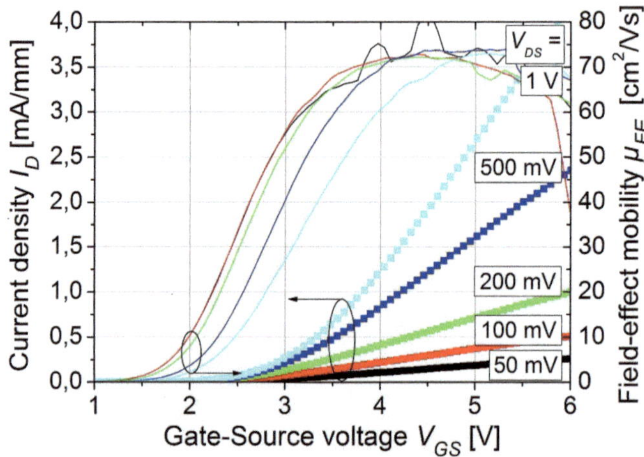

Fig. 5.22: Transfer characteristics and extracted field-effect mobility versus gate-source voltage of a lateral n-channel MOSFET with 12 µm gate length and 100 µm width, according to Fig. 5.21.

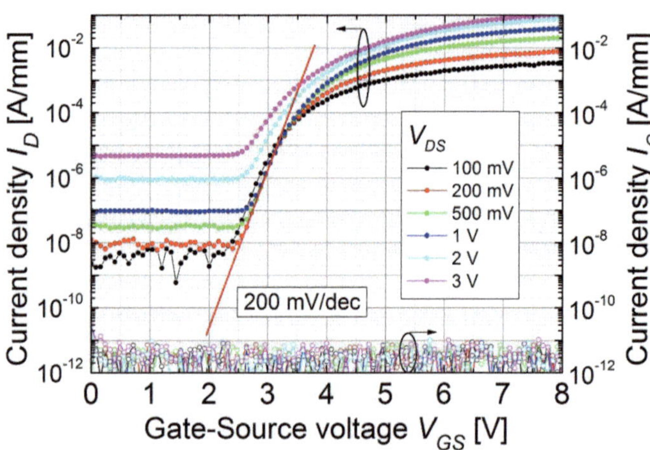

Fig. 5.23: Logarithmic transfer characteristics of a pseudo-vertical MOSFET at different V_{DS}. The same transfer curves as in Fig. 5.17 are shown. Additionally the gate leakage current is plotted. Its value is below the resolution limit of the setup. It does not limit the OFF-state drain leakage current.

Going in the other direction, a doping concentration of $3 \cdot 10^{18}$ cm^{-3} relates to a depletion capacitance of around $1.6 \cdot 10^{-7}$ F/cm^2. Using Equation (2.37) with the body factor of 3.3, an interface trapped capacitance of ($4.2 \cdot 10^{-7}$ Fcm^{-2} or normalized to the elementary charge of $2.6 \cdot 10^{12}$ V^{-1}cm^{-2} can be estimated, which represents an interface trap density of $1.3 \cdot 10^{12}$ cm^{-2} between $V_{GS} = 2$ V and $V_{GS} = 2.5$ V ($D_{it} = 1.3 \cdot 10^{12}$ eV^{-1} cm^{-2}).

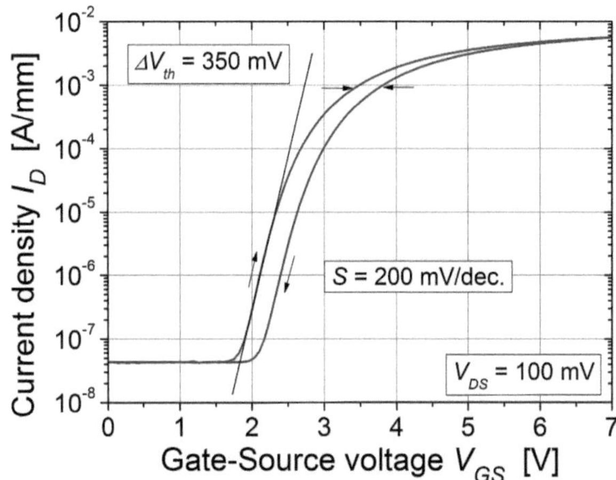

Fig. 5.24: Up and subsequent down sweep of the transfer curve at $V_{DS} = 100$ mV. The shown black line marks the subthreshold fit to obtain the subthreshold slope/swing.

Similar values for S can be obtained on a double-sweep measurement of the transfer characteristics for a similar device, see Fig. 5.24. The hysteresis between the curves reveals negative charge trapping under the gate. The shift of 350 mV corresponds to a negative trap density of $5.6 \cdot 10^{11}$ cm^{-2} located at the high-k/GaN interface. This value is only half of the one previously obtained from the subthreshold slope. It is likely that the hysteresis accesses only rather slow traps deeper in the band gap, due to partial relaxation of fast traps during the back-sweep measurement. The effectively observed trap density is smaller. The subthreshold slope is sensitive to a wider dynamic range of traps, but can only be evaluated in a smaller voltage range.

Micro-structural analysis on the gate MIS-structure was performed by TEM in order to investigate the different interfaces in more detail. In Fig. 5.25 a cross sectional view of one sidewall of the gate trench on a sapphire based pseudo-vertical substrate is shown. The zoomed view (b) shows a high resolution transmission electron micrograph of the vertical gate trench flank close to the upper n$^+$/p GaN interface. Unfortunately a tilt of the etch flank in the sample lamella lead to an inclined view to the high-k/GaN interface. This impedes gatteromg information of the micro-structure of the Al$_2$O$_3$/GaN interface on the atomic scale. Although, the interpretation of the micrograph is difficult, no negative features were observed at the n$^+$/p GaN interface indicating good nucleation during the MBE on the p-GaN MOVPE layer. In part a) of the figure a TEM and SEM of the complete layer stack is shown. The different doping of the layers manifests in the SEM by a varying contrast. Dislocations running through the complete layer stack are visible by TEM imaging. The reduction of dislocations after the nucleation on the substrate during the MOVPE can also be observed.

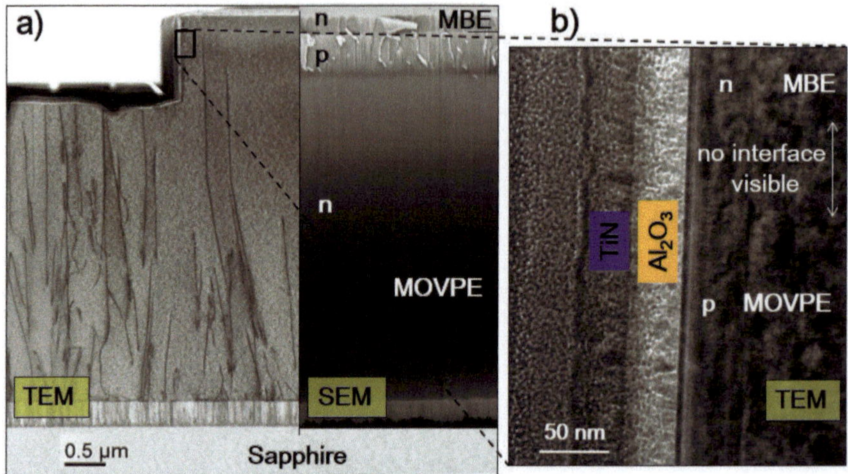

Fig. 5.25: Cross sectional TEM and SEM micrographs (a) taken on a single sidewall of a gate trench on a pseudo-vertical stack and high resolution TEM image of the trench sidewall in vicinity to the upper pn^+-junction (b).

In Section 4.6 the sidewall shape depending on the trench and sidewall orientation is investigated. In the following its influence on the transfer curve of a pseudo-vertical MOSFET will be discussed. In Fig. 5.26 transfer curves of devices with the channel width aligned along a-plane or m-plane are shown. The vertical columnar striped structure from the a-plane oriented sidewall is assumed to be beneficial. Beside small differences in the OFF-state leakage at $V_{GS} < 2$ V mainly the ON-state of the m-plane device is degraded. A reduced subthreshold slope and transconductance is observed, too. Possible reasons for the surprising strong degradation of the m-plane device can be connections problems to the source, which can occur from the horizontal stripes on the m-plane trench sidewall. An effectively increased gate length can additionally contribute. Contrarily, the hysteresis of the double-sweep transfer curve is not degraded, which indicates a similar interface and trapping behaviour. Enhanced trapping at the steps generated during the etching can contribute to degradation. But for both device types a similar hysteresis between forward and backward sweep is observed. Hence, traps with a long relaxation time or fixed charges are considered as unlikely source of the degradation. In contrary, fast trap states cannot be excluded.

Overall it can be stated, that the ON-current, the extracted mobility values, the subthreshold slope and hysteresis presents a promising picture of the processing and of the obtained devices. With further optimization of the process flow and device design, even better values are expected. Mobility values around 150 cm^2/Vs competing to reported values of the OG-FET seem not unrealistic for the here described approach.

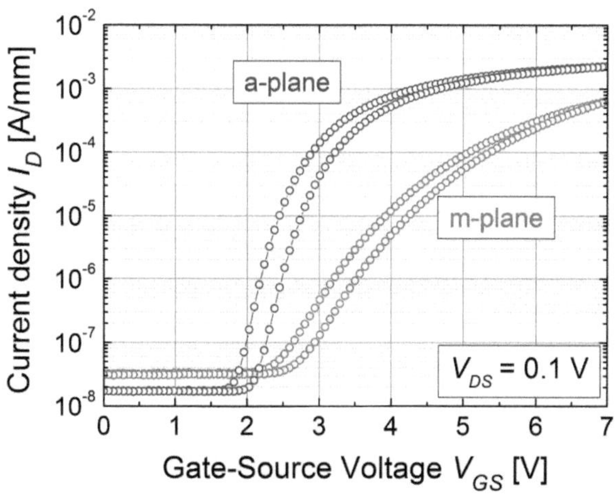

Fig. 5.26: Transfer characteristics for the different oriented trenches after the TMAH etch, referring to Fig. 4.15

One challenge, which was only partially addressed before is the determination of the activation efficiency of the p-GaN. The influence of subsequent processing after the p-GaN growth and its successive activation annealing is additionally related to this . A method to determine the active Mg doping concentration in the body layer in the final MOSFET is presented in the following section.

5.4.2 The body bias effect on the threshold voltage

From the evaluation of the $I(V)$ measurements on TLM structures and $C(V)$ measurements on MIS-structures no clear picture concerning the effective doping concentration of the p-GaN could be obtained. Another method to address the body layer doping is the evaluation of the body bias effect on the threshold voltage. It is shown in the following.
A closer mathematical description of the body bias effect on the threshold voltage is introduced in Section 2.4.6. Transfer characteristics at different body bias were taken at a constant drain-source bias. A positive threshold voltage shift is observed dependent on the body-source voltage. The drain leakage current below $V_{GS} = 2$ V is determined by the body diode reverse leakage, which increases for more negative V_{BS}.
The threshold voltage was evaluated at a current reference value of $1 \cdot 10^{-4}$ A/mm to ensure proper V_{th} extraction also at higher negative V_{BS}. The respective values are shown in Fig. 5.28. With Equation (2.39) the active doping concentration can be calculated. The body control factor can be obtained under use of

$$\gamma = \frac{dV_{th}}{d\sqrt{2\phi_F - V_{BS}}} \tag{5.5}$$

from the measurement. The Fermi potential ϕ_F can be iteratively obtained from the calculated N_{A-} by first assuming a value of 1.6 eV [75]. The calculated doping concentrations from the slope of the threshold voltage versus the body bias is plotted in Fig. 5.28. The value increases at V_{BS} close to 0 V and saturates at more negative V_{BS} to around $2.5 \cdot 10^{17}$ cm^{-3}, which is only around 10 % of the incorporated Mg concentration.

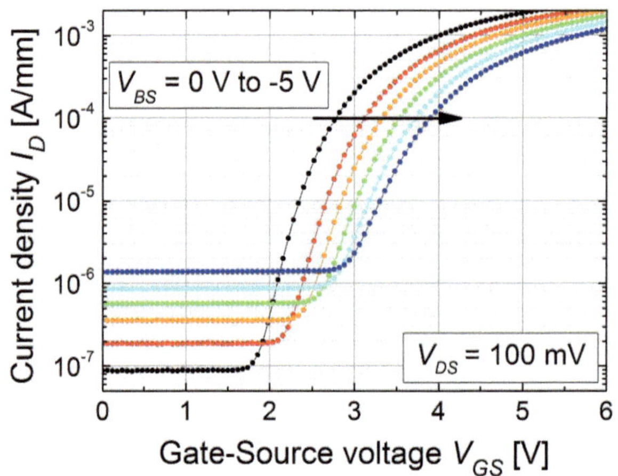

Fig. 5.27: Transfer characteristics for different V_{BS} in steps of 1 V starting at 0 V. A drain-source voltage of 100 mV was applied during the V_{GS} sweep.

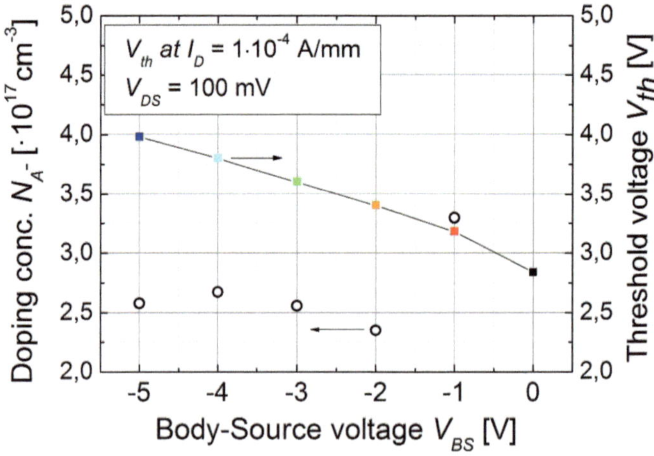

Fig. 5.28: Extracted threshold voltage from the transfer characteristics in Fig. 5.27 at a current of $1 \cdot 10^{-4}$ A/mm and extracted doping concentration.

At this point a short summary is given from the previously discussed results obtained by $I(V)$ characterisation of TLM structures, $C(V)$ measurements on Schottky- (mercury probe) and MIS-structures as well as results extracted from the MOSFET characteristics. On one side, investigations on TLM-structures, which are not overgrown by MBE, indicate almost full activation of the Mg without significant hydrogen passivation. These structures were fabricated after the hydrogen out-diffusion RTA with subsequent shadow mask processing.

On the other side, the MOSFET absolute threshold voltage as well as the body bias effect on the threshold voltage indicate a reduced activation of around 10 - 30 % of the incorporated Mg concentration. In this case, a similar hydrogen out-diffusion anneal was applied before the MBE overgrowth. Incorporation of additional hydrogen during the MBE is excluded, since hydrogen is absent in the growth environment. Therefore the obtained results show a surprising difference. Moreover, the data obtained by $C(V)$ measurements on shadow mask and lithographically processed samples is rather vague and cannot be reliably compared.

Since, several failure sources such as effects from leakage currents, non-ideal ohmic p-GaN contact behaviour and trapping can influence these values a fully conclusive picture is not obtained. Additionally it cannot be excluded, that e.g. during the gate dielectric ALD process additional hydrogen is incorporated. At surface near regions on lateral as well as vertical surfaces a partial re-passivation of the Mg dopant by H can potentially occur.

5.4.3 Device OFF-state

First in this section the OFF-state behaviour of pseudo-vertical devices will be discussed, i.e. the body diode and the MOSFET. Further, turning to the true-vertical MOSFET OFF-state operation, different structural modifications are presented and their influences on the leakage current are discussed. The path for drain, source, body and edge related leakage current contributions are outlined in Fig. 5.29 for a pseudo-vertical MOSFET. The processed pseudo-vertical device has a drift layer thickness of 1 µm and a Si doping of $2 \cdot 10^{16}$ cm^{-3}. The respective leakage characteristics are shown in Fig. 5.30.

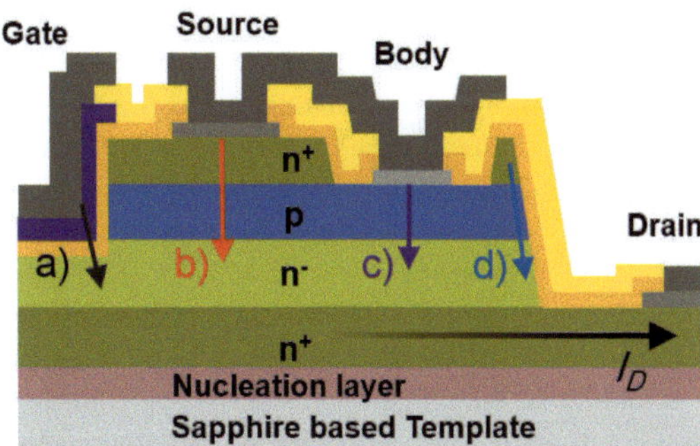

Fig. 5.29: Schematic cross section showing a pseudo-vertical MOSFET with its major leakage path: gate leakage (a), source leakage (b), body leakage/reverse pn-junction leakage (c) and isolation edge leakage (d) which can be composed of body and source leakage contributions.

Additionally, the reverse leakage of a pseudo-vertical pn⁻ diode is shown up to 15 V reverse voltage. The body leakage current of the MOSFET is in the same range as the pn diode leakage current. The deviation of approximately 0.8 orders of magnitude is likely caused by geometric differences between the MOSFETs body diode and the diode test structure. E.g. in the diode test-structure the complete pn junction is covered by a p-GaN contact, whereas the MOSFET body diode is just partially covered by the contact.

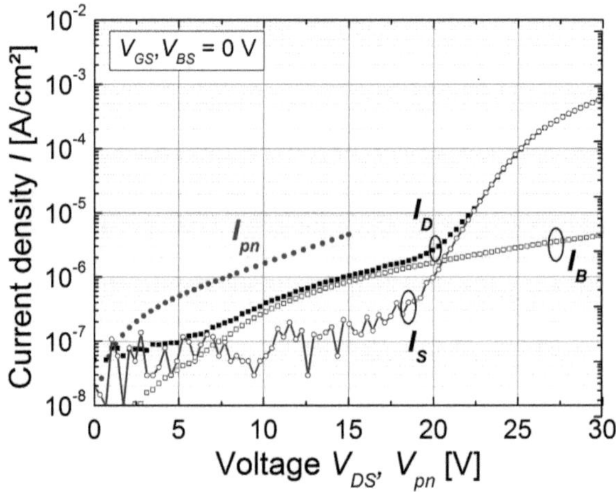

Fig. 5.30: Leakage current characterisation ($I(V_{DS})$) of a pseudo-vertical MOSFET (type B in Fig. 5.31. Additionally the reverse leakage of the pn⁻ diode test structure is included. The gate leakage (not shown) is defined by the lower resolution limit of the measurement setup ($< 1 \cdot 10^{-9}$ A/cm^2).

For the MOSFET two areas in the characteristics can be defined. Up to 20 V the source leakage current is small compared to the body current and the reverse leakage of the pn⁻ junction dominates the drain leakage. In the second area above 20 V, the source leakage strongly increases and limits the lower bounds of the drain leakage. Soft breakdown at around 50 V occurs with the drain leakage running into compliance of the measurement setup. This correlates to a maximum electric field at the p-GaN/n-GaN interface of only 0.7 MV/cm, which is very low compared to the ideal GaN breakdown field of 3.3 MV/cm. Due to the relatively thin drift layer and the high dislocation density in the pseudo-vertical stack very good OFF-state performance is not expected. Thus, in the following focus is put on the true-vertical MOSFET, although the fabricated drift layers with a thickness of 4 µm and a doping concentration of $3 \cdot 10^{16}$ cm^{-3} are still limiting the high voltage performance. In the measurement of the MOSFET OFF-state, it turned out that not only the design of the epitaxial layer stack, but also the configuration of the lateral edge termination has a strong influence on the drain leakage current and breakdown voltage. Further investigations were performed on true-vertical MOSFETs with different edge termination. The respective leakage current versus voltage characteristics in OFF-state (i.e. at $V_{GS} = 0$ V) are shown in figure 5.32. During the measurement of the four terminal MOSFET all current values are continuously monitored. Three different devices were evaluated, which are shown in

Fig. 5.31, and deviate from each other by the structure for the lateral isolation. The respective leakage curves are shown in Fig. 5.32. For a clear visualization the source and gate leakage are not shown, since the gate current contribution can be neglected in these rather low voltage devices and the source current is thus equal to the sum of drain and body leakage current of different polarity.

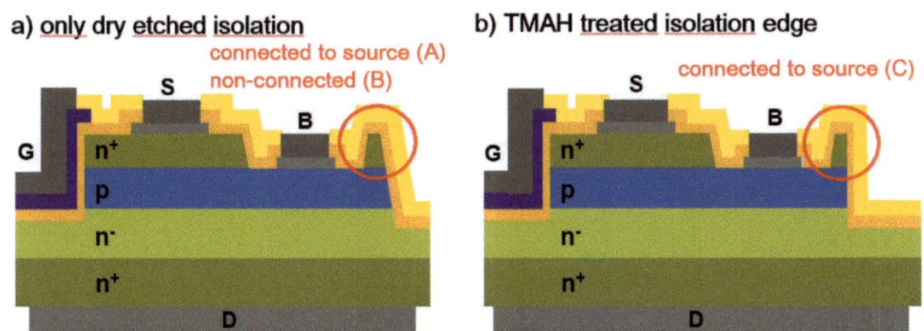

Fig. 5.31: Schematic of the three different device types used for the leakage and breakdown measurements. Device type A and B feature isolation by ICP etching only. Caused by design constrains a small n-GaN island remains, which can be connected to the source (A) or is unconnected (B). For device type C an additional TMAH treatment of the sidewall was conducted similar as for the gate trench. The remaining n-GaN is connected to the source.

In case of device type A the drain current is dominated by a source leakage current with a rather low breakdown voltage of around 90 V. The device isolation was made with an ICP etch only and the remaining n-GaN island close to the isolation edge (depicted with the red circle in Fig. 5.31 is connected to the source contact. An edge-related current originating from this region is likely causing the high source leakage.

In device type B the remaining source layer island at the isolation edge is not connected to the source contact. However, it is also etched by ICP only. In this case the device drain current is bound to the body leakage current, which shows a similar behaviour as for device A and the achieved breakdown voltage is close to the previously found one at around 90 V. This finding indicates an edge related leakage from the body to the drain, which is similar for type A and B. The breakdown is most likely defined by the passivation and vertical isolation edge in these cases.

In contrast, the isolation edge of device type C was additionally treated with TMAH similar to the gate trench. The remaining n-GaN layer is also connected to the source, but in this case the drain leakage current is not dominated by the source current. This indicates, that surface or GaN/passivation interface related leakage currents at the etching flank limit the breakdown voltage when ICP-only isolation is conducted. For TMAH treated surfaces the leakage decreases strongly by more than one order of magnitude compared to device type B and four orders of magnitude compared to device type A, resulting in a breakdown voltage of around 230 V. This corresponds to a critical electric field strength of around 1.8 MV/cm at the p/n^- interface, which is still around 40 % lower compared to the ideal value of GaN of 3.3 MV/cm. Additionally the body leakage current is decreased by one order of magnitude.

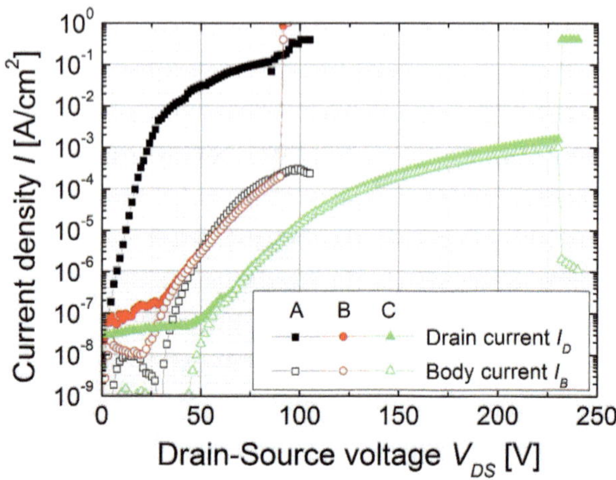

Fig. 5.32: Leakage current scaled to the active area of the MOSFET's drain sided body diode and breakdown voltage measurement ($I(V_{DS})$ at $V_{GS} = 0$ V) of three different devices types (A, B and C). The essential structural difference of the devices is schematically shown in Fig. 5.31.

From the drift layer design with a thickness of 4 µm and a doping of $3 \cdot 10^{16}$ cm^{-3} a higher breakdown voltage is expected. Vertical pn$^-$-diodes with a similar drift layer showed a breakdown voltage of 380 V, as discussed in Section 5.3. The reduction of the breakdown voltage of the MOSFET seems still to be related to the edge termination and/or to the gate module.

The strong decrease of the overall leakage with a wet chemical treatment of the isolation edge gives another hint for the influence of plasma damage on the characteristics of the devices. If a surficial n-type layer is created at the body layer sidewall during dry etching and it is not removed by a TMAH treatment, increased leakage is reasonable.

With a reduction of the doping concentration to values $\leq 1 \cdot 10^{16}$ cm^{-3} and thus a better distribution of the electric field over the drift region an improvement of the breakdown voltage of around 30 % seems possible. However, in this case a slightly increased ON-resistance is expected, too.

6 Summary and conclusion

In this work the fabrication and characterisation of true-vertical and pseudo-vertical MOS-FETs are described and discussed. Detailed characterisation of the device helped to identify improvement possibilities in the processing of various building blocks of the devices. Although, under the context and constrains coming along with the complex overall device integration promising progress could be made. By the investigation of different device modules actual difficulties within the vertical GaN device technology could be worked out. The device OFF-state, the p-type doping of the body layer and the gate module are of special interest. Analyses of the Mg p-type doping has shown that large discrepancies are appearing in various characterisation methods typically used for semiconductor devices in order to validate the incorporated and active doping concentration. Comprehensive investigations are thus necessary to gain reliable information and to reveal a conclusive picture of the layer and device properties.

In order to achieve a broader experimental access without limitations appearing from the availability of bulk GaN substrates, sapphire based GaN templates were initially fabricated to obtain the necessary layer stack. Compromises concerning the leakage behaviour and ON-resistance of the processed devices are necessary. These implications are reasoned by the limited layer thicknesses and higher dislocation density on these substrates. Moreover, the drain contact has to be placed laterally next to the device on the top side of the bottom n^+ layer. The performance of the processed devices is additionally limited by the relaxed design from a lithographical point of view and by constraints appearing from the electrical characterisation. This has to be seen in the context of the relative complex overall device integration, which puts several constraints on the processing options in the process flow. In the following a reasonable extrapolation to an optimized device design is made based on the results obtained on both, pseudo-vertical and true-vertical devices. State of the art devices are compared and an overall evaluation of the technology is given. It also reveals technological limits and further possibilities for advancements.

6.1 Device performance

In this section the presented devices will be classified within the vertical GaN FET technology and compared to conventional silicon devices. SiC devices are additionally listed in order to give a wide overview of various technologies. Previously reported device specifications are summarized in Fig. 6.1. The different device concepts are represented by different research groups all over the world. Silicon device and GaN-HEMT data is shown as reference. Within the different vertical GaN device concepts the MOSFET theoretically combines high power capability and efficient integration without further epitaxial growth on pre-structured surfaces. One challenge is the formation of a reliable gate module, in particular the high-k/GaN interface on the dry and wet chemically etched non-polar trench sidewall. In the following various reports on pseudo- and true-vertical FETs are compared in terms of key device parameters such as breakdown voltage, ON-resistance and threshold voltage.

Contributing with the vertical JFET or Fin power FET the Massachusetts Institute of Technology (MIT) presented a device with very low specific ON-resistance ($R_{ON,sp.}$) of 0.36 mΩcm^2, a breakdown voltage of 800 V and a threshold voltage of 1 V [31]. 1200 V devices were reported in cowork with Singapore-MIT Alliance for Research and Technology and the University of Columbia showing specific ON-resistances of 1 mΩcm^2 and 0.2 mΩcm^2 with similar threshold voltage reaching the power limit for GaN unipolar devices [15, 155].

For the vertical MOSFET technology, Toyoda Gosei achieved device specifications with threshold voltages of 3.5 V and 7 V with specific ON-resistances between 2.2 and 12.1 mΩcm^2 and blocking voltages in the range of 1200 V till 1600 V [67, 156, 157].

A modified MOSFET approach with regrown channel UID layer, called OG-FET was reported by the University of California, Santa Barbara and University of California, Davis (UCSB/UCD) with an $R_{ON,sp.}$ of 1 mΩcm^2 and a blocking voltage of 700 V. The threshold voltage is 1 V. Similar devices where reported with V_{Br} of 900 V and 1435 V and $R_{ON,sp.}$ of 8.4 and 2.2 mΩcm^2, respectively [36, 158].

Addtionally the UCSB demonstrated a 1200 V device with a V_{th} of 1.5 V and a $R_{ON,sp.}$ of 2 mΩcm^2 as well as an OG-FET with a V_{Br} of around 1000 V and a V_{th} of around 3.2 V with an $R_{ON,sp.}$ around 3 mΩcm^2 [159].

A CAVET with a V_{Br} of 1500 V and a $R_{ON,sp.}$ of 2.2 mΩcm^2 was presented in [160] by Avogy. But a rather low threshold voltage of 0.5 V was achieved. Further contribution was made by UCSB and the Arizona State University, Tempe (ASUT) with a low resistive device with an $R_{ON,sp.}$ of 0.4 mΩcm^2 and a V_{Br} of 500 V, but still negative threshold voltage around -7 V [161].

The successful operation of the vertical heterostructure FET (VHFET) was reported by Sumitomo with a 7.6 mΩcm^2 $R_{ON,sp.}$, a breakdown voltage of 672 V and a V_{th} of -1.1 V. A modified VHFET utilizing a p-GaN gate was developed by Panasonic. The device offers superior properties such as a breakdown voltage of 1700 V, a positive threshold voltage of 2.5 V and an $R_{ON,sp.}$ of 1 mΩcm^2 [162, 13] but includes complex growth and processing. A low cost approach for the substitution of the GaN on GaN layer stacks, used for all previously described devices, with GaN on Si templates was demonstrated in [163] from the École Polytechnique Fédérale de Lausanne (EPFL). The presented pseudo-vertical MOSFET has a V_{th} of 6.3 V with an $R_{ON,sp.}$ of 6.8 mΩcm^2 and a V_{Br} of 645 V. However, its drift layer thickness is only 4 µm and the wafer size was 6 inch.

Conclusively it can be seen, that great effort and progress on vertical GaN power FETs has been made in the last decade. The results are very promising for GaN on GaN substrates as well as for the GaN on Si technology. It can be shown that this work fits well to the general development and the obtained devices can compete if properly designed. In this work, large differences for true- and pseudo-vertical devices have been presented, emphasising the importance of high quality substrate material.

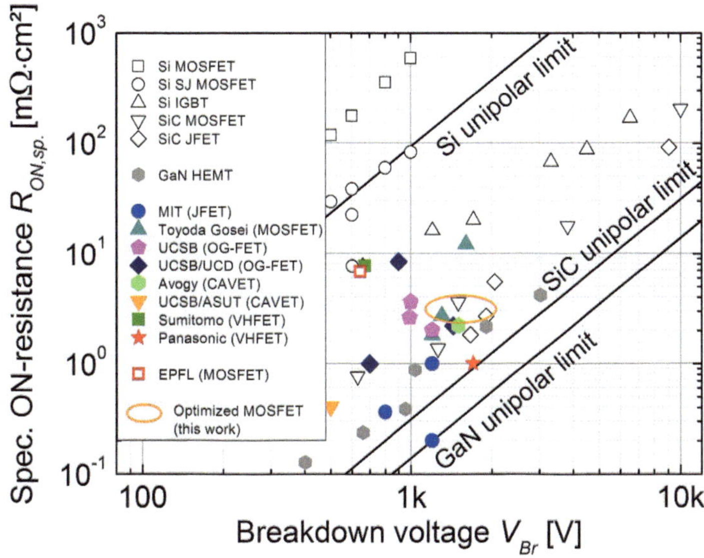

Fig. 6.1: Specific ON-resistance per area versus breakdown voltage of different conventional Si devices, vertical SiC and GaN devices as well as lateral GaN HEMTs.

In chapter 5 the characterisation and discussion of the pseudo- and true-vertical devices was made with special focus on the material properties and characteristics of different building parts of the device. An optimized design suitable for a comparison to previous reported vertical power FETs was not fabricated, but an extrapolation based on the obtained data shall be given in the following. Typical designs for power MOSFETs include a hexagonal structure of the gate module aligned on one facet of the GaN wurtzite crystal, which has a very dense geometry and higher metallization layers in order to improve the power density related to the used surface area. Fig. 6.2 shows a possible schematic cross section and top view of such a device. A gate to gate spacing of 8 µm is considered in this case, which covers 3 µm for the trench, 2 µm for the spacings from the trench towards the source contact and 3 µm for the combined source and body contact. In the subsequent discussion of the MOSFET ON-state the following estimations are applied. The resistive contribution from the substrate side (R_D and R_{Dc}) is neglected due to the large backside area and highly doped substrate in a true vertical design. The source contact is considered with a further reduction of R_{Sc} to 1.5 Ωmm by improving the n^{++}/n$^+$ source layer. The combined source layer resistance and channel resistance is assumed to be around 10 Ωmm with this further improved source contact resistance. The obtained value in Section 5.4.1 is $(4.5 + 8.2) = 12.7$ Ωmm. The area related value of the top part thus can be estimated to 0.8 mΩcm^2. The drift layer resistance with a doping of around $1 \cdot 10^{16}$ cm^{-3} and a thickness of 15 µm contributes with around 2 mΩcm^2. It spends the biggest contribution to the total specific device ON-resistance, which is thus estimated to be about 2.8 mΩcm^2. Typical blocking voltages for these layers are reported to be about 1.5 kV.

With these approximations the described technology together with the optimized design can be compared to previously reported power devices, as shown in Fig. 6.1 by the orange circle, which marks the approximated figure of merit.

Still, a remarkable optimization potential appears comparing most of the reported, as well as the here presented devices to the inherent material limits of GaN. In the following section the major reasons for that shall be discussed.

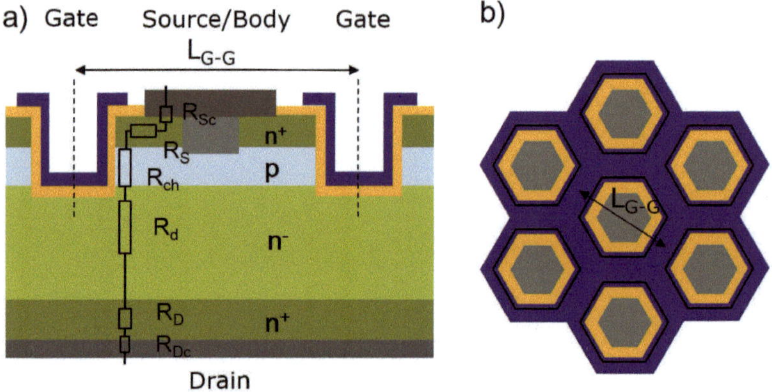

Fig. 6.2: Schematic cross section (a) and a top view (b) for an optimized structure of the vertical MOSFET with a gate to gate spacing (L_{G-G}) of 8 µm

6.2 Current limits of the vertical device technology

In this section major limitation factors of the vertical trench gate MOSFET are discussed. The device has various critical parts, such as the pn-junctions, the gate-sidewall and gate-foot MIS-structures as well as the device isolation and ohmic contacts. These parts shall be considered here. Additionally, material and technological limitations are separately discussed in the following.

- **Magnesium doping and determination of the doping concentration**
  The incorporation of Mg in GaN is well manageable and a wide concentration range from $1 \cdot 10^{17}$ cm^{-3} to $1 \cdot 10^{21}$ cm^{-3} can be achieved. Characterisation by SIMS of the incorporated Mg-concentration is not problematic. In contrary, not straightforward is the determination of the hydrogen content in the layers. Approaches using e.g. deuterium as detection agent are rather impractical for frequent application. The determination of the hydrogen release and thus activation of the magnesium as acceptor mostly has to rely on pure electrical characterisation, which can be hindered by various failure sources during measurement and data extraction. In this respect, ohmic behaviour of metal contacts to lower doped p-GaN is hardly achievable. In particular, contacts on dry etched surfaces without reliable ohmic behaviour complicate the evaluations of the electrical characterisation. Nevertheless, the knowledge about the activation and thus the effective doping is crucial for the prediction of the MOSFETs properties and adjustment of processing parameters.

- **Vertical trench gate formation**
  The formation of a high quality channel and gate trench bottom high-k/GaN interface plays a dominating role for the MOSFET channel mobility, threshold voltage, gate coupling and trapping characteristics. Limitations are currently given by the channel mobility and trapping under the gate. The regrowth of a thin UID GaN layer on the vertical sidewall was reported to be beneficial to increase the channel mobility. However, the regrowth adds another epitaxy step and higher thermal budget during processing. The gate trench with the UID channel layer hat to be processed before the contact formation as part of a gate-first approach. Additionally the threshold voltage of the MOSFET is reduced.

  In general influences of trapping are predominantly reflected by the dynamic behaviour of the device. Associated to the time constant of the traps the turn-ON or turn-OFF transitions are negatively influenced.

- **Geometric and process related limitations**
  One hugh limitation results from difficulties of doping by ion-implantation in GaN. The loss of design freedom is associated to the high crystal damage induced by the ion bombardment. The need for a high thermal budget to recrystallize implant damage limits the flexibility in device structuring for different smaller work packages, e.g. a pocket implant under the gate trench bottom or edge termination structures. Both reduce the peak electric field occurring at the respective device parts.

  Generally, lateral doping profiles can not be easily realized, except by complex etching and regrowth schemes. Instead vertically doping during the growth is performed, which leads to rather simple doping sequences and profiles. Nevertheless, this approach requires very precise control of the doping during growth and high uniformity across the entire wafer.

- **Boundaries by material imperfections and limits**
  Another strong limitation is still given by the GaN material itself, although, recently advanced growth methods and doping procedures have been established. The dislocation density in bulk GaN substrates grown by HVPE is still six orders of magnitude higher than in commercial GaAs or Silicon substrates and two orders of magnitude higher than in SiC substrates. From a feasibility point of view the dislocation density is critical as long as their density is high related to the device area. When the density drops, due to further development to very low values, single small devices are not influenced any more. But still the amount of dislocations could be reflected by the performance yield over the wafer. Additional investigations on how and which dislocations exactly cause leakage and reduced breakdown field strength are necessary. Simultaneously a general development towards a reduction of the dislocation density over larger geometric scales is targeted and ongoing.

  On the other hand the growth of efficient drift layers relies on thick n^- layers with a low unintentional doping concentration. But for MOVPE growth as rule of thumb it can be stated that higher growth rates lead to higher background concentration. It is thus not cost efficient to grow 10 - 15 µm lowly doped material. Further, large foreign substrates demand a good stress and strain management to equalise the lattice mismatch between the GaN epitaxial layers and the substrate material.

  An intrinsic limitation is given by the nature of the direct, wide band-gap material with high recombination coefficient, short carrier lifetime and thus short carrier diffusion length. In power diodes, and IGBTs strong injection is favourable to strongly decrease the drift layer resistance below its value expected from the doping itself (i.e. in low injection conditions). Due to the small diffusion length of the carriers in GaN high injection in a thick drift layer is difficult to achieve. Though, thick drift layers are needed to ensure high reverse voltage capability, elevated on-resistance is created by the thick drift regions. With this stronger boundaries are given for designing power diodes and IGBTs in the future.

6.3 Possibilities for advancements

The here shown work focusses on the fabrication and characterisation of MOSFETs based on pseudo-vertical and true-vertical substrates. Advancements are proposed predominately in this field. Nevertheless, feedback to the growth is most important in order to improve the general properties of the material, which are reflected by the processed devices.
In order to determine the activation efficiency after the incorporation of Mg as p-type dopant a comprehensive investigation of different characterisation methods is necessary. Beside the electrical characterisation, which was also shown in this work, also other physical characterisation methods such as PL can help to obtain a clear picture of the activation of the dopants.

The improvement of the high-k/channel interface is one of the most important objectives for future investigations. On way is the interface engineering of the high-k/GaN interface. By different wet and dry chemical surface treatments as well as post deposition anneals of the dielectric the trapping characteristics and channel mobility can be improved without application of an additional overgrowth. Another way is represented by an additional low defect interface dielectric such as AlN. It can additionally help to improve the interface. Further, anisotropic structuring of the dielectric inside the gate trench can be established. Since, the gate trench corner and bottom MIS-structure has to block high electric fields a thicker dielectric compared to the trench sidewall is beneficial. It ensures that similar gate-coupling and threshold properties are ensured. Additionally the thicker dielectric decreases the gate-drain capacitance and is thus advantageous in dynamic device operation.

The implantation of additional higher doping concentration under metal contacts is usually employed to lower the contact resistance R_c. In particular, for p-GaN, where ohmic behaviour is hardly achievable on layers, which were dry etched before, an additional Mg-implant is interesting. However, high temperature anneals needed after the implant still impede the use of this technology. Similar reasons are following for implants under the gate trench bottom, in order to enhance the breakdown behaviour of the gate or pn junctions as edge termination structures. A Si implant into the upper p-GaN as source region would additionally circumvent the problem of the p-GaN overgrowth with an n-type source layer. Another option to improve the ohmic contact resistance to the p-GaN layer can be the selective overgrowth by MBE. In this case the n-GaN overgrowth on a hard mask patterned wafer can be used to avoid the access etching onto the p-GaN layer. The hard mask material is used as lift-off layer in order to remove GaN residuals grown on it and protects the p-GaN surface during overgrowth. A Si doped GaN layer overgrowth using a Ta patterned wafer is shown in the optical micrograph in Fig. 6.3.

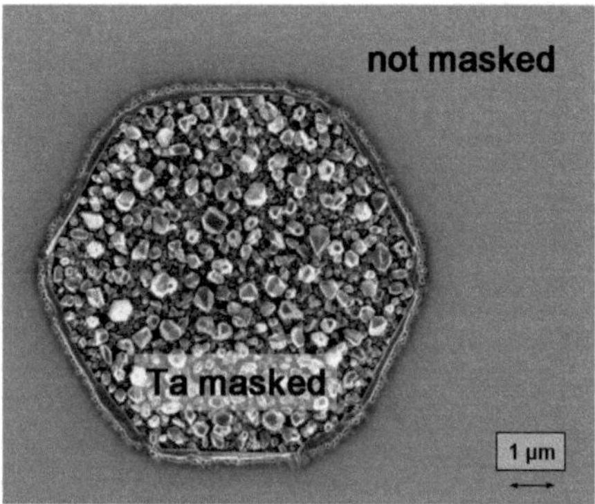

Fig. 6.3: Ta pattered hexagonal structure overgrown by MBE before the removal of the hard mask and the residual GaN on top of it. A smooth surface is achieved next to the hard mask.

A good sticking coefficient during the entire overgrowth procedure including heating, cooling and growth, low partial pressure under the MBE UHV-environment and low diffusion in GaN are beneficial properties for the hard mask material. Additionally good removability is important. Moreover, in case a double layer hard mask is used, the bottom layer can directly be acting as contact material and the upper part as removable lift-off part. In Fig. 6.3 a pure Ta mask was used to show the feasibility of a selective overgrowth process. For further implementations a TaN/Ta hard mask is proposed, which could act as both, mask and contact. In this case only the Ta with the residual GaN has to be removed.

In Section 6.1 a dense device design is proposed, with extrapolated figure of merit. This GaN MOSFET is in competition to current SiC MOS- and JFETs as well as Si IGBTs. Further performance improvements seem still possible by the afore mentioned technical modifications of the concept. However, in terms of yield and material background concentration a well established growth procedure is necessary, which relies on the successive growth on bulk GaN material to decrease the dislocation density. Additionally high purity material with a UID background concentration of $\leq 5{\cdot}10^{15}$ cm^{-3} has to be grown in commercial scale, by a reasonable rate and substrate size.

Bibliography

[1] I. L. Krainsky and V. M. Asnin, "Negative electron affinity mechanism for diamond surfaces," *Applied Physics Letters*, vol. 72, no. 20, pp. 2574–2576, 1998.

[2] M. Benlamri, B. D. Wiltshire, Y. Zhang, N. Mahdi, K. Shankar, and D. W. Barlage, "High breakdown strength schottky diodes made from electrodeposited ZnO for power electronics applications," *ACS Applied Electronic Materials*, vol. 1, no. 1, pp. 13–17, 2019.

[3] U. Oezguer, D. Hofstetter, and H. Morkoc, "Zno devices and applications: A review of current status and future prospects," *Proceedings of the IEEE*, vol. 98, pp. 1255 – 1268, 08 2010.

[4] G. Glover, "The CV characteristics of Schottky barriers on laboratory grown semiconducting diamonds," *Solid-State Electronics*, vol. 16, no. 9, pp. 973–978, 1973.

[5] Y. Zhang, *GaN-based vertical power devices*. Dissertation, Massachusetts Institute of Technology, Department of Electrical Engineering and Computer Science, 2017.

[6] M. Stockmeier, *Realstruktur und Gitterparameter von SiC: Einfluss von Zuchtrichtung, Dotierung und Temperatur*. Dissertation, Friedrich-Alexander-Universität Erlangen-Nürnberg (FAU), 2008.

[7] N. Kaminski and O. Hilt, "Sic and gan devices - competition or coexistence?," in *2012 7th International Conference on Integrated Power Electronics Systems (CIPS)*, pp. 1–11, March 2012.

[8] T. P. Chow and R. Tyagi, "Wide bandgap compound semiconductors for superior high-voltage unipolar power devices," *IEEE Transactions on Electron Devices*, vol. 41, no. 8, pp. 1481–1483, 1994.

[9] B. Ozpineci and L. M. Tolbert, *Comparison of wide-bandgap Semiconductors for Power Electronics applications*. United States. Department of Energy, 2004.

[10] J. Hu, Y. Zhang, M. Sun, D. Piedra, N. Chowdhury, and T. Palacios, "Materials and processing issues in vertical GaN power electronics," *Materials Science in Semiconductor Processing*, vol. 78, pp. 75 – 84, 2018.

[11] H. Amano, Y. Baines, E. Beam, M. Borga, T. Bouchet, P. R. Chalker, M. Charles, K. J. Chen, N. Chowdhury, R. Chu, C. D. Santi, M. M. D. Souza, S. Decoutere, L. D. Cioccio, B. Eckardt, T. Egawa, P. Fay, J. J. Freedsman, L. Guido, O. Häberlen, G. Haynes, T. Heckel, D. Hemakumara, P. Houston, J. Hu, M. Hua, Q. Huang, A. Huang, S. Jiang, H. Kawai, D. Kinzer, M. Kuball, A. Kumar, K. B. Lee, X. Li, D. Marcon, M. März, R. McCarthy, G. Meneghesso, M. Meneghini, E. Morvan, A. Nakajima, E. M. S. Narayanan, S. Oliver, T. Palacios, D. Piedra, M. Plissonnier, R. Reddy, M. Sun, I. Thayne, A. Torres, N. Trivellin, V. Unni, M. J. Uren, M. V. Hove, D. J. Wallis, J. Wang, J. Xie, S. Yagi, S. Yang, C. Youtsey, R. Yu, E. Zanoni,

S. Zeltner, and Y. Zhang, "The 2018 GaN power electronics roadmap," *Journal of Physics D: Applied Physics*, vol. 51, no. 16, p. 163001, 2018.

[12] S. Mandal, A. Agarwal, E. Ahmadi, K. M. Bhat, D. Ji, M. A. Laurent, S. Keller, and S. Chowdhury, "Dispersion free 450 V p GaN-gated CAVETs with Mg-ion implanted blocking layer," *IEEE Electron Device Lett.*, vol. 38, no. 7, pp. 933–936, 2017.

[13] D. Shibata, R. Kajitani, M. Ogawa, K. Tanaka, S. Tamura, T. Hatsuda, M. Ishida, and T. Ueda, "1.7 kV / 1.0 mΩcm^2 normally-off vertical GaN transistor on GaN substrate with regrown p-GaN/AlGaN/GaN semipolar gate structure," in *Electron Devices Meeting (IEDM), 2016 IEEE International*, pp. 10–1, IEEE, 2016.

[14] C. Gupta, S. H. Chan, Y. Enatsu, A. Agarwal, S. Keller, and U. K. Mishra, "OG-FET: An in-situ oxide, GaN lnterlayer-based vertical trench MOSFET," *IEEE Electron Device Letters*, vol. 37, no. 12, pp. 1601–1604, 2016.

[15] Y. Zhang, M. Sun, D. Piedra, J. Hu, Z. Liu, Y. Lin, X. Gao, K. Shepard, and T. Palacios, "1200 V GaN vertical fin power field-effect transistors," in *2017 IEEE International Electron Devices Meeting (IEDM)*, pp. 9.2.1–9.2.4, Dec 2017.

[16] H. Wang, T. Hsieh, Y. Lin, Q. H. Luc, S. Liu, C. Wu, C. F. Dee, B. Y. Majlis, and E. Y. Chang, "AlGaN/GaN MIS-HEMTs with high quality ALD-Al$_2$O$_3$ gate dielectric using water and remote oxygen plasma as oxidants," *IEEE Journal of the Electron Devices Society*, vol. 6, pp. 110–115, 2018.

[17] S. Yagi, M. Shimizu, M. Inada, Y. Yamamoto, G. Piao, H. Okumura, Y. Yano, N. Akutsu, and H. Ohashi, "High breakdown voltage AlGaN/GaN MIS-HEMT with SiN and TiO$_2$ gate insulator," *Solid-State Electronics*, vol. 50, no. 6, pp. 1057 – 1061, 2006. Special Issue: ISDRS 2005.

[18] Z. Tang, Q. Jiang, Y. Lu, S. Huang, S. Yang, X. Tang, and K. J. Chen, "600 V normally off SiN/AlGaN/GaN MIS-HEMT with large gate swing and low current collapse," *IEEE Electron Device Letters*, vol. 34, no. 11, pp. 1373–1375, 2013.

[19] U. K. Mishra, P. Parikh, and Y.-F. Wu, "AlGaN/GaN HEMTs - an overview of device operation and applications," *Proceedings of the IEEE*, vol. 90, no. 6, pp. 1022–1031, 2002.

[20] W. Saito, Y. Takada, M. Kuraguchi, K. Tsuda, and I. Omura, "Recessed-gate structure approach toward normally off high-voltage AlGaN/GaN HEMT for power electronics applications," *IEEE Transactions on electron devices*, vol. 53, no. 2, pp. 356–362, 2006.

[21] Y. Cai, Y. Zhou, K. J. Chen, and K. M. Lau, "High-performance enhancement-mode AlGaN/GaN HEMTs using fluoride-based plasma treatment," *IEEE Electron Device Letters*, vol. 26, no. 7, pp. 435–437, 2005.

[22] Y. Uemoto, M. Hikita, H. Ueno, H. Matsuo, H. Ishida, M. Yanagihara, T. Ueda, T. Tanaka, and D. Ueda, "Gate injection transistor (GIT) - A normally-off Al-GaN/GaN power transistor using conductivity modulation," *IEEE Transactions on Electron Devices*, vol. 54, no. 12, pp. 3393–3399, 2007.

[23] S. Kaneko, M. Kuroda, M. Yanagihara, A. Ikoshi, H. Okita, T. Morita, K. Tanaka, M. Hikita, Y. Uemoto, S. Takahashi, *et al.*, "Current-collapse-free operations up to 850 V by GaN-GIT utilizing hole injection from drain," in *Power Semiconductor Devices & IC's (ISPSD), 2015 IEEE 27th International Symposium on*, pp. 41–44, IEEE, 2015.

[24] G. Meneghesso, G. Verzellesi, R. Pierobon, F. Rampazzo, A. Chini, U. K. Mishra, C. Canali, and E. Zanoni, "Surface-related drain current dispersion effects in AlGaN-GaN HEMTs," *IEEE Transactions on Electron Devices*, vol. 51, no. 10, pp. 1554–1561, 2004.

[25] O. Mitrofanov and M. Manfra, "Mechanisms of gate lag in GaN/AlGaN/GaN high electron mobility transistors," *Superlattices and Microstructures*, vol. 34, no. 1-2, pp. 33–53, 2003.

[26] M. J. Wolter, *Einfluss der Schichteigenschaften auf das elektrische und optoelektrische Verhalten von AlGaN/GaN HEMT Transistoren.* Dissertation, Rheinisch-WestfÃ¤lische Technische Hochschule Aachen, 2004.

[27] S. Karmalkar and U. K. Mishra, "Enhancement of breakdown voltage in AlGaN/GaN high electron mobility transistors using a field plate," *IEEE transactions on electron devices*, vol. 48, no. 8, pp. 1515–1521, 2001.

[28] Y. Dora, A. Chakraborty, L. McCarthy, S. Keller, S. DenBaars, and U. Mishra, "High breakdown voltage achieved on AlGaN/GaN HEMTs with integrated slant field plates," *IEEE Electron Device Letters*, vol. 27, no. 9, pp. 713–715, 2006.

[29] E. A. Jones, F. F. Wang, and D. Costinett, "Review of commercial GaN power devices and GaN-based converter design challenges," *IEEE Journal of Emerging and Selected Topics in Power Electronics*, vol. 4, pp. 707–719, Sept 2016.

[30] D. Ji and S. Chowdhury, "Design of 1.2 kV power switches with low R_{ON} using gan-based vertical JFET," *IEEE Transactions on Electron Devices*, vol. 62, pp. 2571–2578, Aug 2015.

[31] M. Sun, Y. Zhang, X. Gao, and T. Palacios, "High-performance GaN vertical fin power transistors on bulk GaN substrates," *IEEE Electron Device Letters*, vol. 38, pp. 509–512, April 2017.

[32] M. Okada, Y. Saitoh, M. Yokoyama, K. Nakata, S. Yaegassi, K. Katayama, M. Ueno, M. Kiyama, T. Katsuyama, and T. Nakamura, "Novel vertical heterojunction field-effect transistors with re-grown AlGaN/GaN two-dimensional electron gas channels on GaN substrates," *Applied Physics Express*, vol. 3, no. 5, p. 054201, 2010.

[33] S. Yaegassi, M. Okada, Y. Saitou, M. Yokoyama, K. Nakata, K. Katayama, M. Ueno, M. Kiyama, T. Katsuyama, and T. Nakamura, "Vertical heterojunction field-effect transistors utilizing re-grown AlGaN/GaN two-dimensional electron gas channels on GaN substrates," *physica status solidi c*, vol. 8, no. 2, pp. 450–452, 2010.

[34] M. Kanechika, M. Sugimoto, N. Soejima, H. Ueda, O. Ishiguro, M. Kodama, E. Hayashi, K. Itoh, T. Uesugi, and T. Kachi, "A vertical insulated gate AlGaN/GaN heterojunction field-effect transistor," *Japanese Journal of Applied Physics*, vol. 46, no. 6L, p. L503, 2007.

[35] S. Chowdhury and U. K. Mishra, "Lateral and vertical transistors using the AlGaN/GaN heterostructure," *IEEE Transactions on Electron Devices*, vol. 60, pp. 3060–3066, Oct 2013.

[36] D. Ji, C. Gupta, S. H. Chan, A. Agarwal, W. Li, S. Keller, U. K. Mishra, and S. Chowdhury, "Demonstrating ¿ 1.4 kV OG-FET performance with a novel double field-plated geometry and the successful scaling of large-area devices," in *Electron Devices Meeting (IEDM), 2017 IEEE International*, pp. 9–4, IEEE, 2017.

[37] M. Schuster, *Design und Entwicklung einer High-Electron-Mobility Transistor Technologie für Leistungsbauelemente zur Charakterisierung von epitaktisch gewachsenem Galliumnitrid auf 150 mm Si-Wafern*. Dissertation, Technische UniversitÃœt Dresden, 2018.

[38] Y. Zhang, A. Dadgar, and T. Palacios, "Gallium nitride vertical power devices on foreign substrates: a review and outlook," *Journal of Physics D: Applied Physics*, vol. 51, no. 27, p. 273001, 2018.

[39] W. Saito, T. Nitta, Y. Kakiuchi, Y. Saito, K. Tsuda, I. Omura, and M. Yamaguchi, "Suppression of dynamic on-resistance increase and gate charge measurements in high-voltage GaN-HEMTs with optimized field-plate structure," *IEEE transactions on electron devices*, vol. 54, no. 8, pp. 1825–1830, 2007.

[40] J. M. Miller, "Dependence of the input impedance of a three-electrode vacuum tube upon the load in the plate circuit," *Scientific Papers of the Bureau of Standards*, vol. 15, no. 351, pp. 367–385, 1920.

[41] X. Huang, Z. Liu, Q. Li, and F. C. Lee, "Evaluation and application of 600 V GaN HEMT in cascode structure," *IEEE Transactions on Power Electronics*, vol. 29, no. 5, pp. 2453–2461, 2014.

[42] S. M. Sze and K. K. Ng, *Physics of semiconductor devices*. John wiley & sons, 2006.

[43] N. Hizem, A. Fargi, A. Kalboussi, and A. Souifi, "Analysis of the relationship between the kink effect and the indium levels in MOS transistors," *Materials Science and Engineering: B*, vol. 178, no. 20, pp. 1458 – 1463, 2013.

[44] P. Bouillon and T. Skotnicki, "Theoretical analysis of kink effect in C-V characteristics of indium-implanted NMOS capacitors," *IEEE Electron Device Letters*, vol. 19, pp. 19–22, Jan 1998.

[45] S. Nakamura, T. Mukai, M. Senoh, and N. Iwasa, "Thermal annealing effects on p-type Mg-doped GaN films," *Japanese Journal of Applied Physics*, vol. 31, no. 2B, p. L139, 1992.

[46] H. Amano, M. Kito, K. Hiramatsu, and I. Akasaki, "P-type conduction in Mg-doped GaN treated with low-energy electron beam irradiation (LEEBI)," *Japanese Journal of Applied Physics*, vol. 28, no. 12A, p. L2112, 1989.

[47] S. Fischer, C. Wetzel, E. E. Haller, and B. K. Meyer, "On p-type doping in GaN-acceptor binding energies," *Applied Physics Letters*, vol. 67, no. 9, pp. 1298–1300, 1995.

[48] F. Bernardini, V. Fiorentini, and A. Bosin, "Theoretical evidence for efficient p-type doping of GaN using beryllium," *Applied physics letters*, vol. 70, no. 22, pp. 2990–2992, 1997.

[49] J. Lee, S. Pearton, J. Zolper, and R. Stall, "Hydrogen passivation of Ca acceptors in GaN," *Applied physics letters*, vol. 68, no. 15, pp. 2102–2104, 1996.

[50] W. Götz, N. Johnson, J. Walker, D. Bour, H. Amano, and I. Akasaki, "Hydrogen passivation of Mg acceptors in GaN grown by metalorganic chemical vapor deposition," *Applied physics letters*, vol. 67, no. 18, pp. 2666–2668, 1995.

[51] J. Neugebauer and C. G. Van de Walle, "Hydrogen in GaN: Novel aspects of a common impurity," *Physical review letters*, vol. 75, no. 24, p. 4452, 1995.

[52] J. Neugebauer and C. G. Van de Walle, "Role of hydrogen in doping of GaN," *Applied physics letters*, vol. 68, no. 13, pp. 1829–1831, 1996.

[53] W. Götz, N. Johnson, J. Walker, D. Bour, and R. Street, "Activation of acceptors in Mg-doped GaN grown by metalorganic chemical vapor deposition," *Applied Physics Letters*, vol. 68, no. 5, pp. 667–669, 1996.

[54] D. Kim, D. Ryu, N. Bojarczuk, J. Karasinski, S. Guha, S. Lee, and J. Lee, "Thermal activation energies of Mg in GaN: Mg measured by the Hall effect and admittance spectroscopy," *Journal of Applied Physics*, vol. 88, no. 5, pp. 2564–2569, 2000.

[55] T. Tanaka, A. Watanabe, H. Amano, Y. Kobayashi, I. Akasaki, S. Yamazaki, and M. Koike, "P-type conduction in Mg-doped GaN and $Al_{0.08}Ga_{0.92}N$ grown by metalorganic vapor phase epitaxy," *Applied physics letters*, vol. 65, no. 5, pp. 593–594, 1994.

[56] A. K. Viswanath, E.-j. Shin, J. I. Lee, S. Yu, D. Kim, B. Kim, Y. Choi, and C.-H. Hong, "Magnesium acceptor levels in GaN studied by photoluminescence," *Journal of Applied Physics*, vol. 83, no. 4, pp. 2272–2275, 1998.

[57] S. Brochen, J. Brault, S. Chenot, A. Dussaigne, M. Leroux, and B. Damilano, "Dependence of the Mg-related acceptor ionization energy with the acceptor concentration in p-type GaN layers grown by molecular beam epitaxy," *Applied Physics Letters*, vol. 103, no. 3, p. 032102, 2013.

[58] C. G. Van de Walle and J. Neugebauer, "First-principles calculations for defects and impurities: Applications to III-nitrides," *Journal of applied physics*, vol. 95, no. 8, pp. 3851–3879, 2004.

[59] M. A. Reshchikov and H. Morkoc, "Luminescence properties of defects in GaN," *Journal of Applied Physics*, vol. 97, no. 6, p. 061301, 2005.

[60] C. G. Van de Walle, C. Stampfl, and J. Neugebauer, "Theory of doping and defects in III-V nitrides," *Journal of crystal growth*, vol. 189, pp. 505–510, 1998.

[61] Y. Zhang, H.-Y. Wong, M. Sun, S. Joglekar, L. Yu, N. Braga, R. Mickevicius, and T. Palacios, "Design space and origin of off-state leakage in GaN vertical power diodes," in *2015 IEEE International Electron Devices Meeting (IEDM)*, pp. 35–1, IEEE, 2015.

[62] D.-P. Han, C.-H. Oh, H. Kim, J.-I. Shim, K.-S. Kim, and D.-S. Shin, "Conduction mechanisms of leakage currents in InGaN/GaN-based light-emitting diodes," *IEEE Transactions on Electron Devices*, vol. 62, no. 2, pp. 587–592, 2014.

[63] X. Zou, X. Zhang, X. Lu, C. W. Tang, and K. M. Lau, "Fully vertical GaN pin diodes using GaN-on-Si epilayers," *IEEE Electron Device Letters*, vol. 37, no. 5, pp. 636–639, 2016.

[64] C. Gupta, D. Ji, S. H. Chan, A. Agarwal, W. Leach, S. Keller, S. Chowdhury, and U. K. Mishra, "Impact of trench dimensions on the device performance of GaN vertical trench MOSFETs," *IEEE Electron Device Letters*, vol. 38, pp. 1559–1562, Nov 2017.

[65] J. Lutz, *Halbleiter-Leistungsbauelemente: Physik, Eigenschaften, Zuverlässigkeit.* Springer-Verlag, 2012.

[66] M. Weigel, W. Klix, R. Stenzel, R. Hentschel, A. Wachowiak, and T. Mikolajick, "Investigation of vertical GaN-MOSFET breakdown effects by device simulation," in *39th Workshop on Compound Semiconductor Devices and Integrated Circuits, (WOCSDICE)*, 06 2015.

[67] T. Oka, Y. Ueno, T. Ina, and K. Hasegawa, "Vertical GaN-based trench metal oxide semiconductor field-effect transistors on a free-standing GaN substrate with blocking voltage of 1.6 kV," *Applied Physics Express*, vol. 7, no. 2, p. 021002, 2014.

[68] J.-P. Colinge and C. A. Colinge, *Physics of semiconductor devices.* Springer Science & Business Media, 2005.

[69] B. Van Zeghbroeck, "Principles of semiconductor devices," *Colarado University*, vol. 34, 2004.

[70] J. Muth, J. Lee, I. Shmagin, R. Kolbas, H. Casey Jr, B. Keller, U. Mishra, and S. DenBaars, "Absorption coefficient, energy gap, exciton binding energy, and recombination lifetime of GaN obtained from transmission measurements," *Applied Physics Letters*, vol. 71, no. 18, pp. 2572–2574, 1997.

[71] A. Dmitriev and A. Oruzheinikov, "The rate of radiative recombination in the nitride semiconductors and alloys," *Journal of applied physics*, vol. 86, no. 6, pp. 3241–3246, 1999.

[72] W. Gerlach, H. Schlangenotto, and H. Maeder, "On the radiative recombination rate in silicon," *physica status solidi (a)*, vol. 13, no. 1, pp. 277–283, 1972.

[73] A. Galeskas, J. SiCros, V. Grivickas, U. Lindefelt, and C. Hallin, "Proceedings of the 7th International Conference on SiC, III-Nitrides and Related Materials," *Stockholm, Sweden*, pp. 533–536, 1997.

[74] C. C. Enz, F. Krummenacher, and E. A. Vittoz, "An analytical MOS transistor model valid in all regions of operation and dedicated to low-voltage and low-current applications," *Analog integrated circuits and signal processing*, vol. 8, no. 1, pp. 83–114, 1995.

[75] D. K. Schroder, *Semiconductor material and device characterization*. John Wiley & Sons, 2006.

[76] S. Kasap and P. Capper, *Springer handbook of electronic and photonic materials*. Springer, 2017.

[77] M. Miczek, C. Mizue, T. Hashizume, and B. Adamowicz, "Effects of interface states and temperature on the C-V behavior of metal/insulator/AlGaN/GaN heterostructure capacitors," *Journal of Applied Physics*, vol. 103, pp. 104510 – 104510, 06 2008.

[78] A. Winzer, N. Szabo, J. Ocker, R. Hentschel, M. Schuster, F. Schubert, J. Gärtner, A. Wachowiak, and T. Mikolajick, "Detailed analysis of oxide related charges and metal-oxide barriers in terrace etched Al_2O_3 and HfO_2 on AlGaN/GaN heterostructure capacitors," *Journal of Applied Physics*, vol. 118, no. 12, p. 124106, 2015.

[79] C. Mizue, Y. Hori, M. Miczek, and T. Hashizume, "Capacitance-voltage characteristics of Al_2O_3/AlGaN/GaN structures and state density distribution at Al_2O_3/AlGaN interface," *Japanese Journal of Applied Physics*, vol. 50, no. 2R, p. 021001, 2011.

[80] C. M. Jackson, A. R. Arehart, E. Cinkilic, B. McSkimming, J. S. Speck, and S. A. Ringel, "Interface trap characterization of atomic layer deposition Al_2O_3/GaN metal-insulator-semiconductor capacitors using optically and thermally based deep level spectroscopies," *Journal of Applied Physics*, vol. 113, no. 20, p. 204505, 2013.

[81] J. Balland, J. Zielinger, C. Noguet, and M. Tapiero, "Investigation of deep levels in high-resistivity bulk materials by photo-induced current transient spectroscopy. I. Review and analysis of some basic problems," *Journal of Physics D: Applied Physics*, vol. 19, no. 1, p. 57, 1986.

[82] D. Lang, "Deep-level transient spectroscopy: A new method to characterize traps in semiconductors," *Journal of applied physics*, vol. 45, no. 7, pp. 3023–3032, 1974.

[83] R. Hentschel, A. Wachowiak, A. Großer, S. Kotzea, A. Debald, H. Kalisch, A. Vescan, A. Jahn, S. Schmult, and T. Mikolajick, "Extraction of the active acceptor concentration in (pseudo-) vertical GaN MOSFETs using the Body-Bias effect," *Microelectronics Journal*, 2019.

[84] R. Hentschel, A. Wachowiak, A. Großer, A. Jahn, U. Merkel, A. Wille, H. Kalisch, A. Vescan, S. Schmult, and T. Mikolajick, "Pseudo-vertical GaN-based trench gate metal oxide semiconductor field effect transistor," in *11th International Conference on Advanced Semiconductor Devices Microsystems (ASDAM)*, pp. 5–8, Nov 2016.

[85] P. Hofmann, G. Leibiger, M. Krupinski, F. Habel, and T. Mikolajick, "Doping marker layers for ex situ growth characterisation of HVPE gallium nitride," *CrystEngComm*, vol. 19, pp. 788–794, 2017.

[86] P. Hofmann, M. Krupinski, F. Habel, G. Leibiger, B. Weinert, S. Eichler, and T. Mikolajick, "Novel approach for n-type doping of HVPE gallium nitride with germanium," *Journal of Crystal Growth*, vol. 450, pp. 61–65, 2016.

[87] D. Siche and R. Zwierz, "Growth of bulk gan from gas phase," *Crystal Research and Technology*, vol. 53, no. 5, p. 1700224, 2018.

[88] D. Koleske, A. Wickenden, R. Henry, and M. Twigg, "Influence of MOVPE growth conditions on carbon and silicon concentrations in GaN," *Journal of crystal growth*, vol. 242, no. 1-2, pp. 55–69, 2002.

[89] A. Able, *Untersuchungen zur metallorganischen Gasphasenepitaxie von Gruppe III Nitriden auf Silizium (111)*. PhD thesis, University Regensburg, Januar 2005.

[90] S. Fischer, *Über die Herstellung von Gallium- und Aluminiumnitrid aus der Gasphase*. Dissertation, Justus-Liebig-UniversitÃत Gießen, 1999.

[91] A. N. Alexeev, V. P. Chaly, D. M. Krasovitsky, V. V. Mamaev, S. I. Petrov, and V. G. Sidorov, "Features and benefits of III-N growth by ammonia-MBE and plasma assisted MBE," *Journal of Physics: Conference Series*, vol. 541, no. 1, p. 012030, 2014.

[92] D. Katzer, S. Binari, D. Storm, J. Roussos, B. Shanabrook, and E. Glaser, "MBE growth of AlGaN/GaN HEMTs with high power density," *Electronics Letters*, vol. 38, no. 25, p. 1, 2002.

[93] F. Schubert, S. Wirth, F. Zimmermann, J. Heitmann, T. Mikolajick, and S. Schmult, "Growth condition dependence of unintentional oxygen incorporation in epitaxial GaN," *Science and technology of advanced materials*, vol. 17, no. 1, pp. 239–243, 2016.

[94] H. Yamada, K. Iso, M. Saito, K. Fujito, S. P. DenBaars, J. S. Speck, and S. Nakamura, "Impact of substrate miscut on the characteristic of m-plane InGaN/GaN light emitting diodes," *Japanese Journal of Applied Physics*, vol. 46, no. 12L, p. L1117, 2007.

[95] J. Shao, L. Tang, C. Edmunds, G. Gardner, O. Malis, and M. Manfra, "Surface morphology evolution of m-plane (1-100) GaN during molecular beam epitaxy growth: Impact of Ga/N ratio, miscut direction, and growth temperature," *Journal of Applied Physics*, vol. 114, no. 2, p. 023508, 2013.

[96] F. Schulze, A. Dadgar, J. Bläsing, and A. Krost, "Influence of buffer layers on metalorganic vapor phase epitaxy grown GaN on Si (001)," *Applied physics letters*, vol. 84, no. 23, pp. 4747–4749, 2004.

[97] B. Voigtländer, *Scanning probe microscopy*. Springer, 2016.

[98] L. Reimer and S. E. Microscopy, "Physics of image formation and microanalysis," *Springer Ser. in Optical Sciences*, vol. 45, p. 2, 1984.

[99] D. B. Williams and C. B. Carter, "The transmission electron microscope," in *Transmission electron microscopy*, pp. 3–17, Springer, 1996.

[100] W. Burkhardt, *SIMS-Untersuchungen an modernen Halbleitern: GaN und VO₂*. PhD thesis, Universitätsbibliothek Giessen, 1999.

[101] M. Kappers, R. Datta, R. Oliver, F. Rayment, M. Vickers, and C. Humphreys, "Threading dislocation reduction in (0001) GaN thin films using SiNx interlayers," *Journal of Crystal Growth*, vol. 300, no. 1, pp. 70 – 74, 2007. First International Symposium on Growth of Nitrides.

[102] X. H. Wu, L. M. Brown, D. Kapolnek, S. Keller, B. Keller, S. P. DenBaars, and J. S. Speck, "Defect structure of metal-organic chemical vapor deposition-grown epitaxial (0001) GaN/Al₂O₃," *Journal of Applied Physics*, vol. 80, no. 6, pp. 3228–3237, 1996.

[103] T. Hino, S. Tomiya, T. Miyajima, K. Yanashima, S. Hashimoto, and M. Ikeda, "Characterization of threading dislocations in GaN epitaxial layers," *Applied Physics Letters*, vol. 76, no. 23, pp. 3421–3423, 2000.

[104] J. Elsner, R. Jones, P. Sitch, V. Porezag, M. Elstner, T. Frauenheim, M. Heggie, S. Öberg, and P. Briddon, "Theory of threading edge and screw dislocations in GaN," *Physical review letters*, vol. 79, no. 19, p. 3672, 1997.

[105] Y. Xin, S. Pennycook, N. Browning, P. Nellist, S. Sivananthan, F. Omnes, B. Beaumont, J. Faurie, and P. Gibart, "Direct observation of the core structures of threading dislocations in GaN," *Applied physics letters*, vol. 72, no. 21, pp. 2680–2682, 1998.

[106] S. Fischer, *Über die Herstellung von Gallium- und Aluminiumnitrid aus der Gasphase*. PhD thesis, Universitätsbibliothek Giessen, 1999.

[107] V. Potin, P. Ruterana, G. Nouet, R. C. Pond, and H. Morkoç, "Mosaic growth of GaN on (0001) sapphire: A high-resolution electron microscopy and crystallographic study of threading dislocations from low-angle to high-angle grain boundaries," *Phys. Rev. B*, vol. 61, pp. 5587–5599, Feb 2000.

[108] F. K. Yam, L. L. Low, S. A. Oh, and Z. Hassan, "Gallium nitride: an overview of structural defects," in *Optoelectronics-Materials and Techniques*, InTech, 2011.

[109] Z. Fang, B. Claflin, D. C. Look, D. Green, and R. Vetury, "Deep traps in Al-GaN/GaN heterostructures studied by deep level transient spectroscopy: Effect of carbon concentration in GaN buffer layers," *Journal of Applied Physics*, vol. 108, no. 6, p. 063706, 2010.

[110] D. K. Simon, D. Tröger, T. Schenk, I. Dirnstorfer, F. P. G. Fengler, P. M. Jordan, A. Krause, and T. Mikolajick, "Comparative study of ITO and TiN fabricated by low-temperature RF biased sputtering," *Journal of Vacuum Science & Technology A*, vol. 34, no. 2, p. 021503, 2016.

[111] K. Remashan, S. J. Chua, A. Ramam, S. Prakash, and W. Liu, "Inductively coupled plasma etching of GaN using BCl_3/Cl_2 chemistry and photoluminescence studies of the etched samples," *Semiconductor Science and Technology*, vol. 15, no. 4, p. 386, 2000.

[112] S. Pearton, R. J. Shul, and F. Ren, "A review of dry etching of GaN and related materials," *MRS Internet Journal of Nitride Semiconductor Research*, vol. 5, no. 1, p. e11, 2000.

[113] Y. H. Lee, H. S. Kim, G. Y. Yeom, J. W. Lee, M. C. Yoo, and T. I. Kim, "Etch characteristics of GaN using inductively coupled Cl_2/Ar and Cl_2/BCl_3 plasmas," *Journal of Vacuum Science & Technology A*, vol. 16, no. 3, pp. 1478–1482, 1998.

[114] Y. Zhang, M. Sun, Z. Liu, D. Piedra, J. Hu, X. Gao, and T. Palacios, "Trench formation and corner rounding in vertical GaN power devices," *Applied Physics Letters*, vol. 110, no. 19, p. 193506, 2017.

[115] S. D. Burnham, K. Boutros, P. Hashimoto, C. Butler, D. W. Wong, M. Hu, and M. Micovic, "Gate-recessed normally-off GaN-on-Si HEMT using a new O_2-BCl_3 digital etching technique," *physica status solidi c*, vol. 7, no. 7-8, pp. 2010–2012, 2010.

[116] Y. Lin, Y. C. Lin, F. Lumbantoruan, C. F. Dec, B. Y. Majilis, and E. Y. Chang, "A novel digital etch technique for p-GaN Gate HEMT," in *2018 IEEE International Conference on Semiconductor Electronics (ICSE)*, pp. 121–123, Aug 2018.

[117] R. Gutt, L. Kirste, T. Passow, M. Kunzer, K. Köhler, and J. Wagner, "Reduction of the threading edge dislocation density in AlGaN epilayers by GaN nucleation for efficient 350 nm light emitting diodes," *physica status solidi (b)*, vol. 247, no. 7, pp. 1710–1712, 2010.

[118] D. Koleske, M. E. Coltrin, S. R. Lee, and G. Thaler, "Nanostructural engineering of nitride nucleation layers for GaN substrate dislocation reduction," tech. rep., Sandia National Lab.(SNL-NM), Albuquerque, NM (United States), 2007.

[119] C. Seager, S. Myers, A. Wright, D. Koleske, and A. Allerman, "Drift, diffusion, and trapping of hydrogen in p-type GaN," *Journal of applied physics*, vol. 92, no. 12, pp. 7246–7252, 2002.

[120] S. Myers, A. Wright, G. Petersen, W. Wampler, C. Seager, M. Crawford, and J. Han, "Diffusion, release, and uptake of hydrogen in magnesium-doped gallium nitride: Theory and experiment," *Journal of Applied Physics*, vol. 89, no. 6, pp. 3195–3202, 2001.

[121] S. Nakamura, T. Mukai, M. Senoh, and N. Iwasa, "Thermal annealing effects on p-type Mg-doped GaN films," *Japanese Journal of Applied Physics*, vol. 31, no. 2B, p. L139, 1992.

[122] S. Nakamura, N. Iwasa, M. Senoh, and T. Mukai, "Hole compensation mechanism of p-type GaN films," *Japanese Journal of Applied Physics*, vol. 31, no. 5R, p. 1258, 1992.

[123] W. Wampler and S. Myers, "Hydrogen release from magnesium-doped GaN with clean ordered surfaces," *Journal of applied physics*, vol. 94, no. 9, pp. 5682–5687, 2003.

[124] S. Myers and A. Wright, "Theoretical description of H behavior in GaN pn junctions," *Journal of Applied Physics*, vol. 90, no. 11, pp. 5612–5622, 2001.

[125] C. Vartuli, S. Pearton, C. Abernathy, J. MacKenzie, E. Lambers, and J. Zolper, "High temperature surface degradation of III-V nitrides," *Journal of Vacuum Science & Technology B: Microelectronics and Nanometer Structures Processing, Measurement, and Phenomena*, vol. 14, no. 6, pp. 3523–3531, 1996.

[126] S. Pearton, J. Zolper, R. Shul, and F. Ren, "GaN: Processing, defects, and devices," *Journal of applied physics*, vol. 86, no. 1, p. 1, 1999.

[127] W. Li, K. Nomoto, K. Lee, S. M. Islam, Z. Hu, M. Zhu, X. Gao, J. Xie, M. Pilla, D. Jena, and H. G. Xing, "Activation of buried p-gan in MOCVD-regrown vertical structures," *Applied Physics Letters*, vol. 113, no. 6, p. 062105, 2018.

[128] A. Polyakov, N. Smirnov, S. Pearton, F. Ren, B. Theys, F. Jomard, Z. Teukam, V. Dmitriev, A. Nikolaev, A. Usikov, *et al.*, "Fermi level dependence of hydrogen diffusivity in GaN," *Applied Physics Letters*, vol. 79, no. 12, pp. 1834–1836, 2001.

[129] Y. Kuwano, M. Kaga, T. Morita, K. Yamashita, K. Yagi, M. Iwaya, T. Takeuchi, S. Kamiyama, and I. Akasaki, "Lateral hydrogen diffusion at p-GaN layers in nitride-based light emitting diodes with tunnel junctions," *Japanese Journal of Applied Physics*, vol. 52, no. 8S, p. 08JK12, 2013.

[130] R. Hentschel, S. Schmult, A. Wachowiak, A. Großer, J. Gärtner, and T. Mikolajick, "Normally-off operating GaN-based pseudovertical MOSFETs with MBE grown source region," *Journal of Vacuum Science & Technology B, Nanotechnology and Microelectronics: Materials, Processing, Measurement, and Phenomena*, vol. 36, no. 2, p. 02D109, 2018.

[131] J. Hsu, M. Manfra, D. Lang, K. Baldwin, L. Pfeiffer, and R. Molnar, "Surface morphology and electronic properties of dislocations in AlGaN/GaN heterostructures," *Journal of electronic materials*, vol. 30, no. 3, pp. 110–114, 2001.

[132] D. F. Storm, T. O. McConkie, M. T. Hardy, D. S. Katzer, N. Nepal, D. J. Meyer, and D. J. Smith, "Surface preparation of freestanding GaN substrates for homoepitaxial GaN growth by RF-plasma MBE," *Journal of Vacuum Science & Technology B, Nanotechnology and Microelectronics: Materials, Processing, Measurement, and Phenomena*, vol. 35, no. 2, p. 02B109, 2017.

[133] R. Hentschel, J. Gärtner, A. Wachowiak, A. Großer, T. Mikolajick, and S. Schmult, "Surface morphology of AlGaN/GaN heterostructures grown on bulk GaN by MBE," *Journal of Crystal Growth*, vol. 500, pp. 1–4, 2018.

[134] W. Witte, *Building blocks of vertical GaN-based devices.* Dissertation, Rheinisch-Westfälische Technische Hochschule Aachen, 2016.

[135] S. Kotzea, A. Debald, M. Heuken, H. Kalisch, and A. Vescan, "Demonstration of a GaN-based vertical-channel JFET fabricated by selective-area regrowth," *IEEE Transactions on Electron Devices*, vol. 65, pp. 5329–5336, Dec 2018.

[136] P. Hofmann, *Hydride vapour phase epitaxy growth, crystal properties and dopant incorporation in gallium nitride.* Dissertation, Technische Universität Dresden, 2018.

[137] D. Gaskill, L. Rowland, and K. Doverspike, "Electrical properties of AlN, GaN, and AlGaN," *Properties of group III nitrides*, vol. 11, pp. 101–116, 1994.

[138] V. Chin, T. Tansley, and T. Osotchan, "Electron mobilities in gallium, indium, and aluminum nitrides," *Journal of Applied Physics*, vol. 75, no. 11, pp. 7365–7372, 1994.

[139] F. Schwierz, "Low-field electron mobility models for bulk GaN and AlGaN/GaN 2DEGs," in *Extended abstracts of the... Conference on solid state devices and materials*, vol. 2005, pp. 1058–1059, 2005.

[140] M. Tahhan, J. Nedy, S. H. Chan, C. Lund, H. Li, G. Gupta, S. Keller, and U. Mishra, "Optimization of a chlorine-based deep vertical etch of GaN demonstrating low damage and low roughness," *Journal of Vacuum Science & Technology A*, vol. 34, no. 3, p. 031303, 2016.

[141] M. Kodama, M. Sugimoto, E. Hayashi, N. Soejima, O. Ishiguro, M. Kanechika, K. Itoh, H. Ueda, T. Uesugi, and T. Kachi, "GaN-based trench gate metal oxide semiconductor field-effect transistor fabricated with novel wet etching," *Applied physics express*, vol. 1, no. 2, p. 021104, 2008.

[142] X. Cao, S. Pearton, G. Dang, A. Zhang, F. Ren, and J. Van Hove, "GaN n-and p-type Schottky diodes: Effect of dry etch damage," *IEEE Transactions on Electron Devices*, vol. 47, no. 7, pp. 1320–1324, 2000.

[143] C. Caloz and T. Itoh, *Electromagnetic metamaterials: transmission line theory and microwave applications*. John Wiley & Sons, 2005.

[144] N. Papanicolaou, M. Rao, J. Mittereder, and W. Anderson, "Reliable Ti/Al and Ti/Al/Ni/Au ohmic contacts to n-type GaN formed by vacuum annealing," *Journal of Vacuum Science & Technology B: Microelectronics and Nanometer Structures Processing, Measurement, and Phenomena*, vol. 19, no. 1, pp. 261–267, 2001.

[145] A. Davydov, A. Motayed, W. J. Boettinger, R. S. Gates, Q. Xue, H. Lee, and Y. Yoo, "Combinatorial optimization of Ti/Al/Ti/Au ohmic contacts to n-GaN," *physica status solidi (c)*, vol. 2, no. 7, pp. 2551–2554, 2005.

[146] J. S. Kwak, S. Mohney, J.-Y. Lin, and R. Kern, "Low resistance Al/Ti/n-GaN ohmic contacts with improved surface morphology and thermal stability," *Semiconductor science and technology*, vol. 15, no. 7, p. 756, 2000.

[147] Y. Arakawa, K. Ueno, A. Kobayashi, J. Ohta, and H. Fujioka, "High hole mobility p-type GaN with low residual hydrogen concentration prepared by pulsed sputtering," *APL Materials*, vol. 4, p. 086103, 08 2016.

[148] W. Götz, R. Kern, C. Chen, H. Liu, D. Steigerwald, and R. Fletcher, "Hall-effect characterization of III-V nitride semiconductors for high efficiency light emitting diodes," *Materials Science and Engineering: B*, vol. 59, no. 1, pp. 211 – 217, 1999.

[149] B. Podor, "Thermal ionization energy of Mg acceptors in GaN: Effects of doping level and compensation," in *International Conference on Solid State Crystals 2000: Growth, Characterization, and Applications of Single Crystals*, vol. 4412, pp. 299–304, International Society for Optics and Photonics, 2001.

[150] U. Kaufmann, P. Schlotter, H. Obloh, K. Köhler, and M. Maier, "Hole conductivity and compensation in epitaxial GaN: Mg layers," *Physical Review B*, vol. 62, no. 16, p. 10867, 2000.

[151] P. Kozodoy, H. Xing, S. P. DenBaars, U. K. Mishra, A. Saxler, R. Perrin, S. El-hamri, and W. Mitchel, "Heavy doping effects in Mg-doped GaN," *Journal of Applied Physics*, vol. 87, no. 4, pp. 1832–1835, 2000.

[152] S. D. Lester, F. A. Ponce, M. G. Craford, and D. A. Steigerwald, "High dislocation densities in high efficiency GaN-based light-emitting diodes," *Applied Physics Letters*, vol. 66, no. 10, pp. 1249–1251, 1995.

[153] D. Ji, W. Li, and S. Chowdhury, "A study on the impact of channel mobility on switching performance of vertical GaN MOSFETs," *IEEE Transactions on Electron Devices*, vol. 65, pp. 4271–4275, Oct 2018.

[154] H. Otake, K. Chikamatsu, A. Yamaguchi, T. Fujishima, and H. Ohta, "Vertical GaN-based trench gate metal oxide semiconductor field-effect transistors on GaN bulk substrates," *Applied physics express*, vol. 1, no. 1, p. 011105, 2008.

[155] Y. Zhang, M. Sun, J. Perozek, Z. Liu, A. Zubair, D. Piedra, N. Chowdhury, X. Gao, K. Shepard, and T. Palacios, "Large area 1.2 kV GaN vertical power FinFETs with a record switching figure-of-merit," *IEEE Electron Device Letters*, pp. 1–1, 2018.

[156] T. Oka, T. Ina, Y. Ueno, and J. Nishii, "1.8 mΩcm^2 vertical GaN-based trench metal-oxide-semiconductor field-effect transistors on a free-standing GaN substrate for 1.2 kV class operation," *Applied Physics Express*, vol. 8, no. 5, p. 054101, 2015.

[157] T. Oka, T. Ina, Y. Ueno, and J. Nishii, "Over 10 A operation with switching characteristics of 1.2 kV-class vertical GaN trench MOSFETs on a bulk GaN substrate," in *Power Semiconductor Devices and ICs (ISPSD), 2016 28th International Symposium on*, pp. 459–462, IEEE, 2016.

[158] C. Gupta, C. Lund, S. H. Chan, A. Agarwal, J. Liu, Y. Enatsu, S. Keller, and U. K. Mishra, "In situ oxide, GaN interlayer-based vertical trench MOSFET (OG-FET) on bulk GaN substrates," *IEEE Electron Device Letters*, vol. 38, no. 3, pp. 353–355, 2017.

[159] C. Gupta, A. Agarwal, S. H. Chan, O. S. Koksaldi, S. Keller, and U. K. Mishra, "1 kV field plated in-situ oxide, GaN interlayer based vertical trench MOSFET (OG-FET)," in *2017 75th Annual Device Research Conference (DRC)*, pp. 1–2, June 2017.

[160] H. Nie, Q. Diduck, B. Alvarez, A. P. Edwards, B. M. Kayes, M. Zhang, G. Ye, T. Prunty, D. Bour, and I. C. Kizilyalli, "1.5 kV and 2.2 mΩcm^2 vertical GaN Transistors on bulk-GaN substrates," *IEEE Electron Device Letters*, vol. 35, pp. 939–941, Sept 2014.

[161] R. Yeluri, J. Lu, C. A. Hurni, D. A. Browne, S. Chowdhury, S. Keller, J. S. Speck, and U. K. Mishra, "Design, fabrication, and performance analysis of GaN vertical electron transistors with a buried p/n junction," *Applied Physics Letters*, vol. 106, no. 18, p. 183502, 2015.

[162] S. Yaegassi, M. Okada, Y. Saitou, M. Yokoyama, K. Nakata, K. Katayama, M. Ueno, M. Kiyama, T. Katsuyama, and T. Nakamura, "Vertical heterojunction field-effect transistors utilizing re-grown AlGaN/GaN two-dimensional electron gas channels on GaN substrates," *physica status solidi c*, vol. 8, no. 2, pp. 450–452, 2011.

[163] C. Liu, R. A. Khadar, and E. Matioli, "GaN-on-Si quasi-vertical power MOSFETs," *IEEE Electron Device Lett*, vol. 39, no. 1, pp. 71–74, 2018.

A Appendix

A.1 Deduction: Forward diffusion current of the pn-diode

To derive the excess carrier density we start with the deduction in the quasi-neutral n-region at the edge of the space charge region. The charge neutrality equation under non-equilibrium can be written as:

$$n_n p_n = n_i^2 exp\left(\frac{qV_{pn}}{kT}\right).\tag{A.1}$$

Since, $n_n p_n$ account for all majority and minority carriers, we can formulate:

$$n_i^2 exp\left(\frac{qV_{pn}}{kT}\right) = (n_{no} + \delta n_n)(p_{no} + \delta p_n).\tag{A.2}$$

with $\delta n_n = \delta p_n$ as excess and n_{no} and p_{no} as equilibrium carrier densities. If it is assumed that $n_{no} = N_{D+}$ (full ionization, no codoping) and $p_{no} = n_i^2/N_{D+}$ one can obtain the quadratic equation:

$$0 = \delta p_n^2 + \left(N_{D+} + \frac{n_i^2}{N_{D+}}\right)\delta p_n - n_i^2\left(exp\left(\frac{qV_{pn}}{kT}\right) - 1\right),\tag{A.3}$$

which has the solution:

$$\delta p_n = \delta n_n = -\frac{N_{D+} + \frac{n_i^2}{N_{D+}}}{2} \pm \sqrt{\left(\frac{N_{D+} + \frac{n_i^2}{N_{D+}}}{2}\right)^2 + n_i^2\left(exp\left(\frac{qV_{pn}}{kT}\right) - 1\right)}.\tag{A.4}$$

The terms n_i^2/N are usually negligible, especially for wide band gap semiconductors. For the quasi-neutral p-region similar analysis can be performed, which yields:

$$\delta n_p = \delta p_p = -\frac{N_{A-} + \frac{n_i^2}{N_{A-}}}{2} \pm \sqrt{\left(\frac{N_{A-} + \frac{n_i^2}{N_{A-}}}{2}\right)^2 + n_i^2\left(exp\left(\frac{qV_{pn}}{kT}\right) - 1\right)}.\tag{A.5}$$

Further the elevated majority carrier density close to the space charge region leads to an additional voltage drop e.g. V_n in the quasi-neutral n-region. It can be derived from:

$$n_n = n_{no} + \delta n_n = N_{D-} exp\left(\frac{qV_n}{kT}\right)\tag{A.6}$$

and one can get:

$$V_n = kTln\left(1 + \frac{\delta n_n}{n_{no}}\right)\tag{A.7}$$

This voltage is only significant under large excess carrier concentrations i.e. under high injection and is opposite to the applied junction voltage V_{pn}. Thus the depletion layer width has to be corrected as shown in:

$$d_n' = \sqrt{\frac{2\epsilon_r\epsilon_0(V_{Diff} - V_{pn} + V_n)N_{A-}}{qN_{D+}(N_{D+} + N_{A-})}}.\tag{A.8}$$

The diffusion voltage V_{Diff} can be calculated as usual with

$$V_{Diff} = kTln\left(\frac{N_{A-}N_{D+}}{n_i^2}\right). \tag{A.9}$$

An analogous deduction can be performed for the p-type quasi-neutral region, which yields a general equation set to calculate the diffusion current depending on V_{pn} for low and high injection conditions. From these equation it can be seen, that high injection occurs for the low doped region first. At higher voltages high injection occurs additionally in the higher doped region. Generally, due to the voltages V_n and V_p the space charge region doesn't disappear even at higher V_{pn}.

A.2 Deduction: Operation regions in the EKV model

First a short introduction to the taylor series is made, which is given in a general form with:

$$f(x) = f(x_0) + f'(x_0)x + \frac{f''(x_0)}{2!}x^2 + \frac{f'''(x_0)}{3!}x^3... \tag{A.10}$$

Keeping the deduction clear G and D are introduced:

$$G = \frac{q(V_{GS} - V_{th})}{2bkT} \tag{A.11}$$

$$D = \frac{qV_{DS}}{2kT} \tag{A.12}$$

The EKV model is additionally given by:

$$I_D = 2b\mu_{ch}C_{ox}\frac{W_G}{L_G}\left(\frac{kT}{q}\right)^2\left(ln^2\left(1 + exp(G)\right) - ln^2\left(1 + exp(G - D)\right)\right) \tag{A.13}$$

Now the different operation regions are reviewed.

- **Subthreshold region** ($V_{GS} < V_{th}$, $V_{DS} > 0$ V)
  The Taylor series for the term $ln(1 + x)$ at $x_0 = 0$ is:

$$ln(1 + x) = x - \frac{x^2}{2} + \frac{x^3}{3} - \frac{x^4}{4} + ... \tag{A.14}$$

$x_0 = 0$ means similarly $V_{GS} < V_{th}$ so that $exp(G) \to 0$. Setting this into the EKV model (A.13) (up to the second term) one obtains:

$$I_D = 2b\mu_{ch}C_{ox}\frac{W_G}{L_G}\left(\frac{kT}{q}\right)^2\left(\left(exp(G) - \frac{exp(G)^2}{2}...\right)^2 - \left(exp(G - D) - \frac{exp(G-D)^2}{2}...\right)^2\right) \tag{A.15}$$

It can be seen, that the linear term dominates in this case, since $exp(G)$ is small. With $exp(G)^2 = exp(2G)$ the following, simplified equation is obtained:

$$I_D = 2b\mu_{ch}C_{ox}\frac{W_G}{L_G}\left(\frac{kT}{q}\right)^2\left(exp(2G) - exp(2(G - D))\right) \tag{A.16}$$

... which can be rearranged to:

$$I_D = 2b\mu_{ch}C_{ox}\frac{W_G}{L_G}\left(\frac{kT}{q}\right)^2\left(1 - exp(-D)\right)exp(2G) \tag{A.17}$$

The term $exp(-D)$ can be considered as negligible since D is always positive and the exponential term much smaller than unity. To extract the subthreshold swing the logarithm has to be drawn:

$$log(I_D) = log\left(2b\mu_{ch}C_{ox}\frac{W_G}{L_G}\left(\frac{kT}{q}\right)^2\right) + log(e)2G \qquad (A.18)$$

...and the derivation after dV_{GS} with substituted G is:

$$\frac{dlog(I_D)}{dV_{GS}} = log(e)\frac{q}{bkT} \qquad (A.19)$$

With $log(e)^{-1} = ln(10)$ and substituted body factor b this finally results in:

$$\frac{dV_{GS}}{dlog(I_D)} = ln(10)\frac{kT}{q}\left(1 + \frac{C_D}{C_{ox}}\right) \qquad (A.20)$$

... which is the typical expression for the subthreshold swing without interface traps.

- **Linear region** $(V_{GS} > V_{th},\ V_{DS} < V_{GS} - V_{th})$
  Starting from the EKV model (A.13) the term related to G and D are much larger than unity and one obtains:

$$I_D = 2b\mu_{ch}C_{ox}\frac{W_G}{L_G}\left(\frac{kT}{q}\right)^2\left(G^2 - (G-D)^2\right) \qquad (A.21)$$

With the second binomial formula $(G-D)^2 = G^2 - 2GD + D^2$:

$$I_D = 2b\mu_{ch}C_{ox}\frac{W_G}{L_G}\left(\frac{kT}{q}\right)^2\left(2GD - D^2\right) \qquad (A.22)$$

Which is with the respective substitutions for G and D:

$$I_D = \mu_{ch}C_{ox}\frac{W_G}{L_G}\left((V_{GS} - V_{th})V_{DS} - \frac{1}{2}bV_{DS}^2\right) \qquad (A.23)$$

- **Quadratic region** $(V_{GS} > V_{th},\ V_{DS} > V_{GS} - V_{th})$
  Again starting from the EKV model (A.13) the term related to G is much larger than unity, the second logarithm can be neglected due to $G - D < 0$ and one obtains:

$$I_D = 2b\mu_{ch}C_{ox}\frac{W_G}{L_G}\left(\frac{kT}{q}\right)^2\left(G^2\right) \qquad (A.24)$$

Which finally results in:

$$I_D = \mu_{ch}C_{ox}\frac{W_G}{L_G}\left(\frac{(V_{GS} - V_{th})^2}{2b}\right) \qquad (A.25)$$

Figures

Tables

Abbreviations

Abbreviation	Explanation
AC	alternating current
AFM	atomic force microscope / microscopy
ALD	atomic layer deposition
AlGaN	aluminium gallium nitride
AlN	aluminium nitride
Al_2O_3	aluminium oxide
ASUT	Arizona state university, Tempe
BCl_3	bortrichloride
BF	bright field
C	carbon or diamond
CAVET	current apperture vertical transistor
CBM	conduction band minimum
Cl_2	molecular chlorine
Cp_2Mg	bis(cyclopentadienyl)magnesium
CVD	chemical vapour deposition
D	drain electrode
D_c	cascode drain electrode
DC	direct current
DCS	dichlorsilane
DF	dark field
DUT	device under test
EPFL	École Polytechnique Fédérale de Lausanne
FET	field effect transistor
FP	field plate
G	gate electrode
G_c	cascode gate electrode
GaAs	gallium arsenide
GaCl	gallium chloride
GaN	gallium nitride
GIT	gate injection transistor
GND	ground
H^+	proton / ionized atomic hydrogen
H_2	molecular hydrogen
HEMT	high electron mobility transistor
HCl	hydrochloric acid
HF	hydrofluoric acid
HFET	heterostructure field effect transistor
HR-TEM	high-resolution transmission electron microscopy
HVPE	hydrid vapour phase epitaxy
ICP	inductively coupled plasma
IGBT	insulated gate bipolar transistor

## Abbreviation	Explanation

Abbreviation	Explanation
JFET	junction field effect transistor
LED	light emitting diode
MBE	molecular beam epitaxy
MIS	metal-insulator-semiconductor
MIT	massechusetts institute of technology
MOS	metal-oxide-semiconductor
MOVPE	metal organic vapour phase epitaxy
NH_3	ammonia
N_2	molecular nitrogen
OG-FET	overgrown channel field effect transistor
O_2	molecular oxygen
O_3	ozone
PECVD	plasma enhanced - chemical vapour deposition
PL	photoluminescence
PVD	physical vapor deposition
RF	radio frequency
RMS	root mean square
RTA	rapid thermal annealing
S	source electrode
S_c	cascode source electrode
SC1	standard clean 1
SE	secondary electron
SEM	scanning electron microscopy
SF_6	sulfurhexafluorine
Si	silicon
SI	semi-insulating
SiC	silicon carbide
SiH_4	silane
Si_3N_4	silicon nitride
SIMS	secondary ion mass spectroscopy
SRH	Schockley Read Hall
STEM	scanning transmission electron microscopy
Ta	tantalum
TaN	tantalum nitride
TDD	threading dislocation density
TED	threading edge dislocation
TEM	transmission electron microscopy
TiN	titanium nitride
$TiCl_4$	titanium(IV)-chloride
TLM	transfer line method
TMAH	tetramethylammonium hydroxide
TMAl	trimethylaluminium
TMD	threading mixed type dislocation

Abbreviation **Explanation**

Abbreviation	Explanation
TMGa	trimethylgallium
TSD	threading screw dislocation
UHV	ultra high vacuum
UCD	university of Califonia, Davis
UCSB	university of Califonia, Santa Barbara
VBM	valence band maximum
VFET	vertical field effect transistor
VHFET	vertical heterostructure field effect transistor
VPE	vapour phase epitaxy
2DEG	two dimensional electron gas
4H-SiC	fourfold layer sequence hexagonal SiC polytype
6H-SiC	sixfold layer sequence hexagonal SiC polytype
3C-SiC	threefold layer sequence cubic SiC polytype

Symbols

Symbol	Unit	Explanation
A	cm^2	area
A_A		acceptor activation efficiency
b		body capacitance factor
C	F/cm^2	capacitance per area
C_D	F	depletion layer capacitance
C_{DS}	F	drain-source capacitance
C_{GD}	F	gate-drain capacitance
C_{GS}	F	gate-source capacitance
C_{it}	F	interface trap capacitance
C_{ox}	F/cm^2	oxide capacitance
C_s	F	series capacitance
C_p	F	parallel capacitance
C_{waf}	\$/cm^2	wafer and epitaxy cost per area
D_f		duty cycle
D		dissipation factor
d	μm	depth (vertical, z-direction)
d_{drift}	μm	thickness of the drift layer
d_{MIS}	μm	MIS-structure diameter
d_n	μm	n$^+$-GaN depletion depth
d'_n	μm	depletion depth in triangular case
d_{ox}	μm	oxide thickness
d_p	μm	thickness of the p-GaN depletion layer
d_{p-GaN}	μm	thickness of the p-GaN layer
d_{pn}	μm	pn-diode diameter
d_{Sub}	inch	substrate diameter
E_A	eV	concentration dependent acceptor activation energy
E_{A0}	eV	zero acceptor activation energy
E_{Br}	V/m	electric breakdown field strength
E_C	eV	conduction band energy
E_D	eV	donor state energy
E_F	eV	Fermi energy
$E_{F(n^+)}$	eV	Fermi energy in n$^+$-GaN
$E_{F(p)}$	eV	Fermi energy in p-GaN
E_g	eV	band gap energy
E_{max}	eV	maximum electric field
E_{ox}	eV	oxide electric field
E_V	eV	valence band energy
f	Hz	frequency
f_s	Hz	switching frequency
G		gate overdrive factor
g_e		degeneracy factor for electrons, 2

Symbol	Unit	Explanation
g_h		degeneracy factor for holes, 4
g_m	A/Vmm	transconductance
h	nm	AFM scan height
I	A	current
I_B	A/mm	body current density
I_D	A/mm	drain current density
I_{Don}	A/mm	drain ON-current density
I_{Doff}	A/mm	drain OFF-current density
I_G	A/mm	gate current density
I_{Gon}	A/mm	gate ON-current density
I_{Goff}	A/mm	gate OFF-current density
I_{pn}	A/cm^2	pn-diode current density
I_{Si}	A	Silicon source supply current
I_{TLM}	A/mm	TLM current density
k	eV/K	Boltzmann constant, $8.617 \cdot 10^{-5}$ eV/K
L	μm	breadth or length
L_D	cm	debye length
L_G	μm	gate length
L_{G-G}	μm	gate to gate spacing
n_{if}		ideality factor
n	cm^{-3}	electron concentration
n_i	cm^{-3}	intrinsic carrier concentration
N	cm^{-3}	doping concentration
N_A	cm^{-3}	acceptor concentration
N_{A-}	cm^{-3}	ionized acceptor concentration
N_{A-T}	cm^{-3}	ionized acceptor concentration calculated by a temperature dependent mobility
N_C	cm^{-3}	conduction band density of states, $4.3 \cdot 10^{15}\, T^{(3/2)}$ cm^{-3}
N_{CV}	cm^{-3}	apparent doping concentration from $C(V)$ measurement
N_D	cm^{-3}	donor concentration
N_{D+}	cm^{-3}	ionized donor concentration
N_{D+n+}	cm^{-3}	ionized donor concentration in n$^+$-GaN
N_{D+n-}	cm^{-3}	ionized donor concentration in n$^-$-GaN
N_{Mg-H}	cm^{-3}	hydrogen passivated magnesium concentration
N_{Mgi}	cm^{-3}	interstitial magnesium concentration
N_{Mg0}	cm^{-3}	incorporated magnesium concentration
N_V	cm^{-3}	valence band density of states, $8.9 \cdot 10^{15}\, T^{(3/2)}$ cm^{-3}
P_D	W	dynamic power loss
p	cm^{-3}	hole concentration
P_S	W	static power loss
q	C	elementary charge, $1.602 \cdot 10^{-19}$ As
Q_{eff}	C	effective interface charge
Q_{FB}	C	flat band charge
Q_{ISS}	C	input charge

Symbol	Unit	Explanation
Q_{OSS}	C	output charge
Q_{Mg}	sccm	magnesium precursor flow
R_c	Ωmm	contact resistance
R_{ch}	Ωmm	channel resistance
R_d	Ωmm	drift layer resistance
R_D	Ωmm	drain layer resistance
R_{Dc}	Ωmm	drain contact resistance
R_{DSon}	Ωmm	device ON-resistance
R_{DS}	Ωmm	drain-source resistance
R'_{DS}	Ωmm	extrinsic drain-source resistance
R_{sh}	Ω/sq	sheet resistance of a layer
R_{Msh}	Ω/sq	metal line sheet resistance
$R_{ON,sp}$	Ωcm^2	specific device ON-resistance
R_p	Ω	parallel resistance
R_{RMS}	nm	root mean square roughness
R_s	Ω	series resistance
R_S	Ωmm	source layer resistance
R_{Sc}	Ωmm	source contact resistance
R_{scr}	Ωmm	resistance accompanied to the space charge region
R_{sh}	Ω/sq	sheet resistance
R_{TLM}	Ωmm	measured two point resistance on TLM structures
S	mV/decade	subthreshold swing
T	K or °C	temperature
u	cm	circumference
V	V	voltage
V_{BS}	V	body-source voltage
V_{Br}	V	device breakdown voltage
V_{DB}	V	drain-body voltage
V_{DG}	V	drain-gate voltage
V_{Diff}	V	diffusion voltage
V_{DS}	V	drain-source voltage
$V_{DS,ch}$	V	intrinsic channel drain-source voltage
V_{DSoff}	V	drain-source OFF-state voltage
V_{FB}	V	flat band voltage
V_{GD}	V	gate-drain voltage
V_{GS}	V	gate-source voltage
$V_{GS,ch}$	V	intrinsic channel gate-source voltage
V_{GSoff}	V	gate-source OFF-state voltage
V_n	V	voltage over the quasi-neutral n-region
V_{pn}	V	pn-diode voltage
V_p	V	voltage over the quasi-neutral p-region
v_{sat}	cm/s	electron saturation velocity
V_{th}	V	threshold voltage
V_{TLM}	V	TLM voltage

Symbol	Unit	Explanation
W	μm	width
W_G	μm	gate width
x	μm	lateral x-direction
y	μm	lateral y-direction
z	μm	profiling depth, z-direction
Z_0	Ω	characteristic transmission line impedance
α	meVcm	activation energy coefficient
γ	$\sqrt{\text{V}}$	body control factor
ϵ_r		relative permittivity
ε_{GaN}		GaN relative permittivity
ε_{ox}		oxide relative permittivity
ε_0	As/Vcm	vacuum permittivity, $8.854 \cdot 10^{-14}$ As/Vm
κ	W/Km	thermal conductivity
μ_{ch}	cm^2/Vs	channel mobility
μ_{eff}	cm^2/Vs	effective channel mobility
μ_{FE}	cm^2/Vs	field effect channel mobility
μ_n	cm^2/Vs	electron mobility
μ_p	cm^2/Vs	hole mobility
$\mu_p(T)$	cm^2/Vs	temperature dependent hole mobility
ρ_c	Ωcm^2	specific contact resistance
ρ_{GaN}	Ωcm	specific material resistance of GaN
Φ_{MS}	eV	metal-semiconductor work function difference
ϕ_F	V	Fermi potential with respect to the intrinsic Fermi level
ϕ_S	V	surface potential
χ	eV	electron affinity

Postamble & Acknowledgement

Die vorliegende Arbeit entstand an der NaMLab gGmbH im Zeitraum von April 2014 bis Ende 2019. Ursprünglich hatte ich für die experimentelle Phase der Arbeit, in meiner damals noch jugentlichen Naivität, mit 3 Jahren gerechnet. Dieser Zeitraum war auch für das "ZweiGaN" Projekt geplant, was mich während der ersten Jahre meiner Doktorandenzeit in sehr motivierender Hinsicht begleitete. Mit Hilfe dieses Projektes wurde durch das gesamte GaN Team die, für uns zu dieser Zeit neue, vertikale GaN Technologie aus dem NaMLab-boden gestampft. Nun habe ich mich nicht nur wissenschaftlich weiterentwickelt, sondern auch erkannt, dass ich verschiedenen Teilgebieten der Arbeit mehr Aufmerksamkeit und Geduld entgegenbringen will. Der Grund dafür liegt nicht nur an den technologischen Schwierigkeiten sondern auch an den hervorragenden Arbeitsbedingungen, die durch das Zusammenwirken von wissenschaftlichem und ingeneurtechnischem Verständnis/Vorgehen am NaMLab und IHM der TU-Dresden zustande kommen. Mit der dadurch möglichen fortschreitenden Entwicklung, die unter anderem durch Folgeprojekte wie dem "Enkrist" ermöglicht wurde, hoffe ich, dass ich mit dieser Arbeit einen sinnvollen, wenn auch bescheidenen Beitrag zur Entwicklung einer nachhaltigen Technologie liefern konnte. Der Erfahrschatz, den ich durch meine Arbeit am NaMLab erlangen konnte, erstreckt sich dabei weit über die hier vorgestellte Arbeit.

Ausgehend davon möchte ich an dieser Stelle meinen Dank an Dr. Alexander Ruf und Prof. Thomas Mikolajick richten. Durch die Möglichkeit der Promotion am NaMLab und der TU-Dresden haben sie mir ein einzigartiges Forschungsumfeld mit den entsprechenden Ressourcen zur Verfügung gestellt.

Mindestens gleichwertig ist mein Wunsch Dr. Andre Wachowiak zu danken. Er hat mich während meiner gesamten Doktorandenzeit tatkräftig, gewissenhaft und mit großem Engagement und Erfahrungschatz als direkter Ansprechpartner für jegliche Herrausforderung unterstützt.

Ebenso möchte ich die Unterstützung weiterer Kollegen, Mitstreiter und ehemaliger Doktoranden, darunter Dr. S. Schmult, Dr. M. Schuster, Dr. A. Winzer, Dr. F. Schubert, Dr. A. Krause, F. Winkler und A. Calzolaro würdigen. Sie haben mir besonders in meiner Anfangzeit aber auch im weiteren Verlauf meiner Arbeit mit vielen fruchtbaren Diskussion als auch mit einem stets offenem Ohr zur Seite gestanden. Besonderer Dank gilt auch C. Richter, A. Jahn, Dr. U. Merkel, N. Szabó, J. Gärtner, A. Großer und T. Neuhaus, diejenigen Techniker und Mitarbeiter, von denen ich stets bei praktischen Problemen als auch moralisch unterstützt wurde.

Weiterhin würde ich gern dem übrigen NaMLab Kollegium, der GaN Gruppe der RWTH Aachen, den Kollegen von FCM und der HTW Dresden sowie allen Personen danken, die am Zustandekommen der Arbeit teil hatten.

Ein kleiner aber vielleicht der wichtigste Personenkreis kann an dieser Stelle keinen Dank für ein fachliches Zutun zu dieser Arbeit für sich beanspruchen. Dafür aber umso mehr durch familiären Rückhalt, Freundschaft, die unentwegte Motivation und nicht zuletzt wegen meiner "work-life balance".